Aviator's
Guide to GPS

Aviator's Guide to GPS

Bill Clarke

TAB Books
Division of McGraw-Hill, Inc.
New York   San Francisco   Washington, D.C.   Auckland   Bogotá
Caracas   Lisbon   London   Madrid   Mexico City   Milan
Montreal   New Delhi   San Juan   Singapore
Sydney   Tokyo   Toronto

FIRST EDITION
FIRST PRINTING

© 1994 by **TAB Books**.
TAB Books is a division of McGraw-Hill, Inc.

Library of Congress Cataloging-in-Publication Data
Clarke, Bill (Charles W.)
    Avaitor's guide to GPS / by Bill Clarke
        p.   cm.
    Includes index.
    ISBN 0-07-011271-1 (H)   ISBN 0-07-011272-X (pbk.)
    1. Navigation (Aeronautics) 2. Global Positioning System.
I. Title.
TL696.A77C58   1994
629.135'1—dc20   93-44964
                                                                   CIP

Acquisitions Editor: Jeff Worsinger
Editorial Team: Charles Spence, Editor
            Susan W. Kagey, Managing Editor
            Joanne Slike, Executive Editor
Production Team: Katherine G. Brown, Director
            Lorie L. White, Proofreader
            Terry W. Hite, Systems Manager
Design Team: Jaclyn J. Boone, Designer
            Brian Allison, Associate Designer   AV1
Cover design and illustration: Sandra Blair, Harrisburg, Pa.   011272X

OTHER BOOKS BY THE AUTHOR

The Cessna 150 & 152—2nd Edition
The Cessna 172—2nd Edition
The Illustrated Buyers' Guide to Used Airplanes—3rd Edition
The Pilots' Guide to Affordable Classics—2nd Edition
The Piper Indians

This book is dedicated to the first navigators,
those unknown souls who ventured forth following the stars.

Notices

Contents

3 Civilian aviation use   38

4 GPS applications   44

Acknowledgments

THIS BOOK WAS MADE possible by the cooperation and assistance of the following:

Karen Behring, Project Representative Corporate Communications, Jeppesen Sanderson
Charles "Chip" Boyd, Owner, AeroNautical Products
Deborah Brown, Vice President, Technology Solutions
Jonathon Cassat, Manager Corp. Communications, Garmin International
Kit Ehrlich, Director Marketing Communications, Micrologic
Michael Feldbauer, Vice President, J&W Marketing Associates, Inc.
Kenneth Foret, Manager Sales & Marketing, Northstar Technologies, Inc.
Richard Fractor, President, Eventide, Inc.
Kenneth Gebhart, President, Celestaire, Inc.
Terri Germinario, Marketing Communications Manager, Eventide, Inc.
Jim Glodfelty, Manager of Dealer Sales, S-Tec Corp.
Richard Groux, Marketing Manager, Trimble Navigation
Mike Gvili, President, ZYCOM Corp.
Susan Hamner, Vice President Marketing/Sales, ARNAV Systems, Inc.
J. Hume, Vice President, Magnavox Electronic Systems, Inc.
Gordon Kaiser, President, OACCQPOINT
Howie Keefe, President, Air Chart Systems
Sally Kenvin, Corporate Communications Manager, Ashtech, Inc.
Marilyn Lovelady, Art Director, Terra Avionics
Brian Mahoney, Satellite Program Office of the FAA
Patrick Millard, Vice President Marketing, NARCO Avionics, Inc.
Willena Nistico, Marketing Communications, II Morrow, Inc.
Michael Richmond, Captain, United States Air Force
Mark Rubenstein, Marketing & Sales Manager, Magellan Systems Corp.
Harold Sanford, Director of Sales & Marketing, MEMTEC, Corp.
Lyne Sikkel, Advertising Administrator, AlliedSignal, Inc.
Jim Thebeau, Manager of Marketing, Rockwell International

Barbara Thomas, Public Relations, Trimble Navigation
Tom Tishauser, Product Manager, Motorola, Inc.
Almarah Uwzayaz, First Lieutenant, United States Air Force
James White, Public Relations Manager, Magellan Systems Corp.
Donald Wilson, President, Peacock Systems, Inc.
Todd Winter, Vice President, Mid-Continent Instrument Co., Inc.

    A special thanks to the research librarians of the Voorheesville Public Library,
of Voorheesville, New York—the finest library a writer could ask for.

Introduction

MANY YEARS AGO a young Indian brave asked Chief Two Owls, "Where are we?" The chief grunted and tersely said, "Here! We are here." When the brave asked where they had been the previous day, the chief replied, "There!"

It is the relationship between *here* and *there* that forms the basis of, and need for, navigation. Navigation is the means used to determine where *here* is, where *there* is, and how to get *there* from *here*.

Nature has provided an internal means of accurate navigation to migratory birds and spawning fish, but not so for mankind. Man, out of necessity, developed and applied navigation as a learning experience.

Navigation can be as simple as monitoring visual signs, such as the positions of natural or manmade landmarks. For the aviator this means pilotage, the art of looking out the window to see where you are in reference to mountains, streams, roads, bridges, tall buildings, lakes, and other landmarks. This is fine for those roaming about an area providing such ready clues, but suppose you desire independence from visual contact. During flights over remote areas, night operations, IFR, and just to know exactly where you are, the need exists for a simple and accurate locating of *here*.

For the past 40 years, serious air navigation has consisted primarily of various forms of radio direction finding in the guise of NDBs (nondirectional radiobeacons) and the VOR (very high frequency omnidirectional range). Of recent importance is the LORAN-C system, primarily designed for marine use and later adapted for aviation use. Each system has drawbacks and all suffer from a general lack of consistently good accuracy. Then, in the 1970s, a system of global radio navigation was envisioned by the United States armed services to be based upon satellites. The now-operational program, known as the NAVSTAR Global Positioning System, can locate an airplane accurately within a few feet in three dimensions: latitude, longitude, and altitude.

To utilize the GPS system the user has only to operate a specialized receiver and directly read the location from the device's digital display. Air navigation has become, through the availability of GPS, so simple a child can do it. And to an accuracy that exceeds that to which most maps can be read.

This book is written to initiate those interested in GPS-based navigation into the history and theory of the system. Discussions include the hardware user receivers and government-provided space-based and land-based equipment needed to utilize the system. Accuracy and applications will be explained, including civilian and military purposes and the national defense security measures currently in place that protect the system.

As a basis for navigation usage, short treatises are written about alternate forms of radio navigation: NDBs, the current VOR System, LORAN-C, and others. History, strong points, and weaknesses of each will be addressed. Where GPS will take navigation for the general-aviation pilot and what projects the FAA is currently working on, and their plans for the future, will be reviewed.

The basics of GPS receivers are explained, including what options are available and what they will do for the user. The reader will learn a new vocabulary of terms and their definitions. GPS navigation for the pilot will be thoroughly covered from entry-level devices through integrated equipment using LORAN, moving maps, and more. Examples of use are given in a hands-on approach. Finally, an overview of all the commonly available current general-aviation-related GPS equipment is displayed, with manufacturers' comments and specifications.

In all, this book is written to familiarize the general-aviation pilot/airplane owner with the NAVSTAR Global Positioning System in an easy-to-understand manner, allowing later purchase and use of equipment suited to meet current and planned needs. Additionally, it provides background materials and resource information for those desiring further study of GPS.

1

GPS from inception

THE NAVSTAR Global Positioning System, hereinafter referred to simply as GPS, is a space-based radio positioning network providing properly equipped users with highly accurate PVT (position, velocity, and time) information. When fully operational, GPS will furnish continuous worldwide service.

The GPS was developed by the U.S. Department of Defense as a satellite-based radionavigation system to be the DOD's primary means of radionavigation well into the next century. The motivation for the system's development was to meet the common radio positioning and navigation needs of a broad spectrum of users, military and civil, with a single system capable of diversified applications versus a collection of expensive, unshared systems of limited applications and usage. GPS is designed to serve an unlimited number of users anywhere on the ground, at sea, in the air, and in near space.

Projected GPS users

The DOD expects extensive use of GPS in almost every military mission area. The U.S. Department of Transportation (DOT), and others, are evaluating the use and potential applications of GPS to meet civil navigation requirements.

The U.S. has encouraged NATO (North Atlantic Treaty Organization) participation in the development and deployment of GPS military user equipment. At this time Belgium, Canada, Denmark, France, Germany, Italy, the Netherlands, Norway, Spain, the United Kingdom, and Australia are participating in the development of such equipment. The DOD also has working relationships with other friendly nations and is sharing information that is designed to create further interest in military employment of the system. When GPS becomes operational, the DOD plans to phase out its remaining common-use radionavigation systems.

Because of the accuracy, worldwide coverage, and flexibility provided by GPS, widespread national and international civil use of the system is anticipated.

The overall system provides:

- Accurate 3D (three-dimensional) determination of position, velocity, and time
- Passive operation
- All-weather operation
- Real-time positioning
- Continuous operation
- Usable in a hostile environment (military uses)

It is important to note that GPS receivers operate passively (no communication with the satellite system), therefore, a limitless number of simultaneous users can exist.

In accordance with the Federal Radionavigation Plan (FRP) jointly prepared by the Department of Defense and Department of Transportation:

> . . . many existing navigation systems are under consideration for replacement by GPS beginning in the mid- to late-1990s. GPS may ultimately supplant less-accurate systems such as LORAN-C, Omega, VOR, DME, TACAN, and Transit, thereby substantially reducing federal maintenance and operating costs associated with these current radionavigation systems.

National security caveat

In the interest of U.S. national security, the highly accurate and dependable GPS has built-in features which can deny accurate service to unauthorized users, prevent spoofing (passing of incorrect data meant to deceive users), and reduce receiver susceptibility to jamming.

These security measures, designed only with the military in mind, can cause considerable difficulties for unauthorized users. Essentially, an unauthorized user is defined as anyone without a specific military need and/or mission.

GPS PROGRAM HISTORY

Since the early 1960s the U.S. Air Force (USAF) and U.S. Navy (USN) have operated or studied assorted satellite navigation systems. The navy sponsored two programs, Transit and Timation.

Transit: First operational in 1964, Transit is currently providing surface navigation service for ships.

Timation: A high-tech research program for a two-dimensional (latitude and longitude) navigation system.

During the same period of time, the air force conducted concept studies assessing a three-dimensional (latitude, longitude, and altitude) navigation system called 621B.

GPS program management

In 1973 the U.S. Deputy Secretary of Defense directed that the air force be the executive service to consolidate the Timation and 621B programs into a single, all-weather navigation system to be called the NAVSTAR Global Positioning System.

The NAVSTAR GPS Joint Program Office (JPO) was established in July 1973 at U.S. Air Force Systems Command/Space and Missile Systems Organization (SAMSO), Los Angeles AFB, California. The JPO is staffed by personnel from the USAF, USN, U.S. Army (USA), U.S. Marine Corps (USMC), U.S. Coast Guard (USCG), U.S. Defense Mapping Agency (DMA), NATO nations, and Australia.

Development phases of GPS

By December of 1973 the JPO had received approval to start the concept validation phase (Phase One) of the GPS program. This phase included concept studies, projected system performance, and feasibility. Phase One was completed in 1979.

Phase Two was subsequently started and included full-scale equipment development (including the development of GPS user equipment) and system testing. That phase ended in 1985.

The third phase (Phase Three) started in 1985, with the production of GPS equipment and further system developments leading to the completed satellite constellation, Master Control Station (MCS), and advanced user equipment.

Operational capability

The term FOC (full operational capability) defines the condition when full and supportable military capability is provided by a system. GPS FOC will be declared by the Secretary of Defense when 24 operational satellites (Block II/IIA types) are in their assigned orbits and when the constellation has successfully completed testing. Three of the 24 satellites will be orbiting spares that can easily be moved to replace a faulty satellite.

An Initial Operational Capability (IOC) was attained when 24 GPS satellites (Block I/II/IIA types) were operating in their assigned orbits, available for navigation use, and providing service. This total included three operational spares in orbit.

Notification of IOC came from the Secretary of Defense following an assessment by the USAF (the system operator) that the constellation could sustain required levels of accuracy and availability throughout the IOC period. IOC occurred on December 8, 1993. Full military FOC is expected in 1995.

Prior to IOC, GPS was considered to be in the process of development for operational purposes, therefore signal availability and accuracy were subject to change.

Operation and logistical support

Starting in 1986, overall operation of the Control and Space Segments of GPS was managed by the USAF 2nd Space Wing at Falcon AFB, Colorado. Prior to that time operation was from a prototype master control station operated from Vandenberg AFB, California.

Fig. 1-1. Falcon AFB, Colorado, showing the master control station.

In January 1992 the U.S. Air Force activated the 2nd Operations Squadron (2 SOPS), 50th Space Wing, at Falcon AFB, Colorado, with an assigned mission to operate the master control station (Fig. 1-1).

The costs

The GPS is expensive from the point of view of the United States taxpayer. It is installed, operated, and maintained for all to use, on a worldwide basis, with no user charge.

The overall initial costs to the taxpayer for GPS has been in excess of $10 billion. Annually, additional funds must be expended for system upkeep. The tentative GPS budget for 1994 amounts to more than $500 million for the DOD and various smaller sums for DOT, the USCG, and the FAA.

GPS SEGMENTS

GPS consists of space, control, and user segments. Each segment has specific duties and responsibilities (Fig. 1-2):

- Space—satellites
- Control—ground-based tracking and system adjustment
- User—receiver/processor

Space segment

The NAVSTAR space segment is a constellation (group) of GPS satellites in semi-synchronous orbits around the earth.

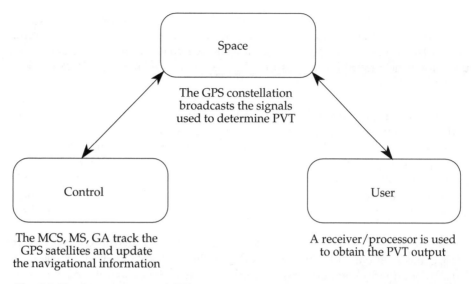

Space

The GPS constellation
broadcasts the signals
used to determine PVT

Control

User

The MCS, MS, GA track the
GPS satellites and update
the navigational information

A receiver/processor is used
to obtain the PVT output

Fig. 1-2. The three segments of GPS.

When fully operational, as currently envisioned, the constellation will ensure continual system availability. The entire contingent of GPS satellites are in six orbital planes, with three or four operational satellites in each plane.

The orbital planes of the satellites have an inclination of 55 degrees (relative to the equator) and an altitude of 20,200 km (10,900 miles). They typically complete an orbit in about 12 hours. Specifically, there is a four minute per day difference between a satellite's orbit time and the rotation of the earth. The satellites are positioned in such a manner that a minimum of five are normally observable (in view) by a user anywhere on earth at any given time. The radio broadcasts from the satellites are called signals-in-space (SIS) (Fig. 1-3).

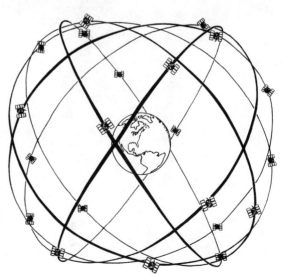

Fig. 1-3. The deployed
constellation. 24 satellites (3 are
orbiting spares. Repeating
ground tracks of 23 hours and
56 minutes. 5 satellites normally
in view. Based on U.S. Government Publication
YEE-82-009D illustration

Space vehicle history. The first NAVSTAR SVs (space vehicles), designated as Navigation Technology Satellite (NTS) Block Zero numbers 1 and 2, were refurbished Timation satellites built by the Naval Research Lab (NRL). SVs 1 and 2 were used for concept validation purposes and carried the first atomic clocks ever launched into space.

These experimental SVs functioned only for short periods of time, however, proved the concept of time-based ranging using spread-spectrum radio signals and precise time derived from orbiting atomic clocks.

Block I SVs. The Block I SVs were built by Rockwell International and launched between 1978 and 1985 from Vandenberg AFB, California. They supported most of the system's testing program (Fig. 1-4). Of the eleven Block I SVs launched, one was lost as the result of a launch failure, others have failed due to deterioration of their atomic clocks or suffered failures of their attitude control system.

Many of the Block I SVs remained in orbit and continued in GPS service far longer than originally planned. For example, Block I SVN (satellite vehicle number) 9, continued to operate reliably for nearly twice its design life of five years. Improvements made to the atomic clocks on later Block I SVs may allow them to operate into the 21st century.

Block II SVs. Block II SVs, also built by Rockwell International, physically differ from Block I SVs in shape and weight (Fig. 1-5) and incorporate design differences affecting security and integrity. Significant Block II SV enhancements include:

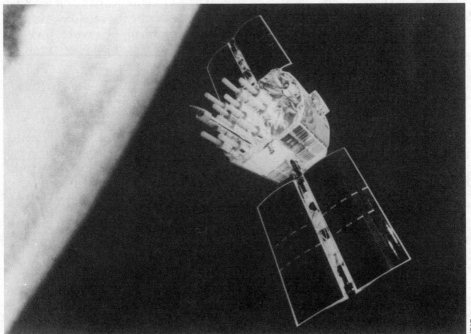

USAF

Fig. 1-4. Block I satellite.

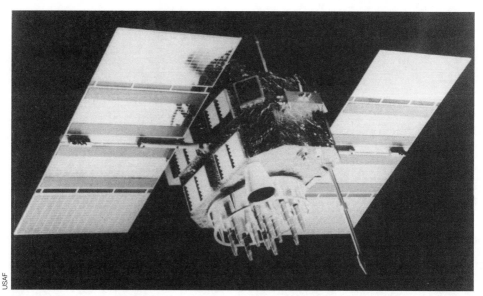

USAF

Fig. 1-5. Block II satellite.

- Radiation-hardened electronics to improve reliability and survivability
- Full selective availability (SA) and anti-spoofing (A-S) capabilities to provide for system security
- Automatic detection of certain error conditions and switching to nonstandard code transmission or default NAV-msg (navigation of the satellite) data to maximize system integrity

The first Block II SV, SVN 14, was launched in February 1989 from Cape Canaveral AFS, Florida, using a Delta II MLV (medium launch vehicle) and became operational for global use in April 1989 (Fig. 1-6). Block II SVs were originally designed for shuttle deployment, however, after the Challenger disaster the decision was made to launch via the Delta II.

SV numbers 13 through 21 are straight Block II type satellites, however, Block II satellites numbers 22 and up have the additional capability of operating for up to 180 days without contact from the control segment and are called Block IIA types (3.5 days limit on older type Block II SVs).

Block IIR SVs. Phase III of the space segment began in the middle of 1989 with the procurement of 20 additional satellites (Block IIR SVs), for replacement purposes, from Martin Marietta (formerly General Electric) (Fig. 1-7).

Block IIR SVs provide an identical signals-in-space interface to the user as the Block II SVs, however, under a survivability scenario the operations of Block IIR SVs differ from the Block I and Block II SVs. Specifically, in the event the control segment cannot contact Block I and Block II SVs, the satellites will simply continue to transmit the stored NAV-msg data previously uploaded by the control segment for 3.5 and 180 days respectively. The improved Block IIR SVs, however, are capable of AUTONAV (autonomous navigation) and can generate their own NAV-msg data. This

Fig. 1-6. Launch of a Block II SV on a Delta II MLV.

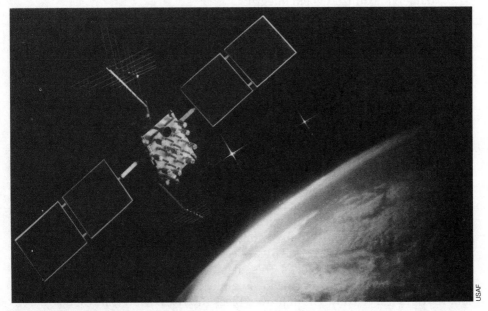

Fig. 1-7. Block IIR satellite.

capability enables Block IIR SVs to maintain full system accuracy without control segment support. Block IIR SVs are scheduled to start launching in the middle 1990s.

Block IIF satellites will follow the Block IIR versions. There is no further information about these units available at this time.

Satellite launches

The oldest currently working (at the time of this writing) Block I satellite was launched in 1984. Subsequent launches have been as follows (in date order of launch):

SVN	PRN	Launched	Set operational
Block I Satellites			
9	13	6-13-84	7-19-84
10	12	9-8-84	10-3-84
11	3	10-9-85	10-30-85
Block II Satellites			
14	14	2-14-89	4-15-89
13	2	6-10-89	8-10-89
16	16	8-18-89	10-14-89
19	19	10-21-89	11-14-89
17	17	12-22-89	1-11-90
18	18	1-24-90	2-14-90
20	20	3-26-90	4-18-90
21	21	8-2-90	8-31-90
15	15	10-1-90	10-15-90
Block IIA Satellites			
23	23	11-26-90	12-10-90
24	24	7-4-91	8-30-91
25	25	2-23-92	3-24-92
28	28	4-10-92	4-25-92
26	26	7-7-92	7-23-92
27	27	9-9-92	9-30-92
32	1	11-22-92	12-11-92
29	29	12-28-92	1-5-93
22	22	2-3-93	4-4-93
31	31	3-30-93	4-13-93
37	7	5-13-93	6-12-93
39	9	6-26-93	7-20-93
35	5	8-30-93	9-28-93
34	4	10-26-93	11-29-93

NOTE: SVN refers to the satellite number and PRN indicates the pseudorandom noise code of the satellite.

Control segment

The control segment consists of a master control station at Falcon AFB in Colorado Springs, Colorado, and five monitor stations (MS) situated at Hawaii, Kwajalein, Diego Garcia, Ascension, and collocated with MCS at Falcon AFB (Figs. 1-8 and 1-9).

The MCS is the central processing facility for the network and is manned 24 hours per day, 7 days per week. It is tasked with tracking, monitoring, and managing the GPS satellite constellation and for updating the navigation data messages.

The task of the monitor stations is to passively track all GPS satellites in view

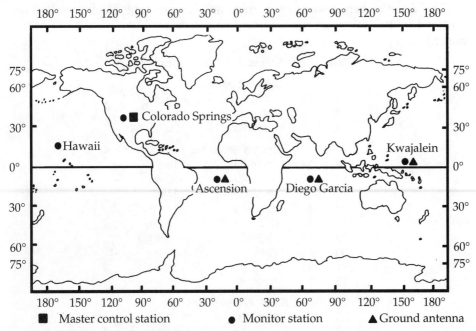

Fig. 1-8. Locations of control segment installations. U.S. Government Publication YEE-82-009D

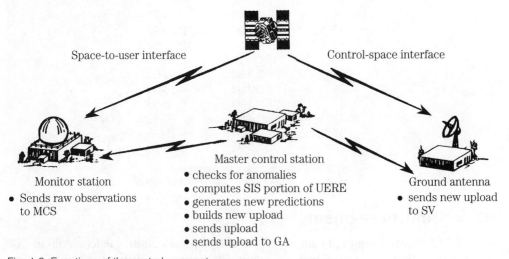

Fig. 1-9. Functions of the control segment. U.S. Government Publication YEE-82-009D

(up to eleven simultaneously) and collect ranging data from each. The MSs are very accurate radio receivers located at precisely surveyed locations.

Monitor stations do little data processing themselves, rather, send their raw measurements and NAV-msg observations to the master control station. The information is processed by the MCS, where satellite ephemeris and clock parameters are estimated and predicted. Using this information, the MCS periodically uploads the ephemeris and clock data to each satellite for retransmission in the NAV-msg.

The updated information is transmitted to the satellites via the ground antennas (GAs), which are also used for transmitting and receiving satellite control information. All the monitoring stations, except Hawaii and Falcon, are equipped with GAs to communicate with the satellites (Fig. 1-10).

The user segment

The user segment is comprised of a variety of military and civilian receiver/processors specifically designed to receive, decode, and process the GPS satellite ranging codes and navigation data messages (Figs. 1-11 and 1-12). These in-

USAF

Fig. 1-10. Ground antenna station in Hawaii.

Fig. 1-11. A general-aviation pilot might use this Apollo 2001 GPS receiver.

II Morrow

Fig. 1-12. Military user of the GPS.

clude stand-alone units and integrated equipment using GPS in combination with other navigation systems (i.e., LORAN, Omega, or Inertial Navigation Systems).

Manufacturers of the receiver/processor (sometimes referred to as a GPS receiver or UE for user equipment) have designed equipment to track the GPS satellite radio signals and provide astonishingly accurate (as compared to other current systems) position, velocity, and time information. Generalized applications include:

- Navigation
- Positioning
- Time transfer
- Surveying

Due to the wide potential for specialized and varied applications of GPS, user equipment can vary significantly in design and function. However, on the civilian user market the differences, except for geodesy (surveying) use, are not greatly pronounced, mainly amounting to calculation and database features (bells and whistles). For example, the primary difference between a hand-held GPS receiver intended for civilian boating and one intended for general aviation is the inclusion of an aviation facility database in the latter.

2

Theory of GPS operation

GPS PROVIDES USERS with three-dimensional position, velocity determination, and precision time transfer. It does not provide a navigation capability per se. It is a positioning system upon which various means of navigation may be based.

Unlike true radio navigation systems, GPS provides position and velocity information that is determined with respect to the World Geodetic System 1984 (WGS-84) map datum absolute earth coordinates in latitude and longitude and not with respect to a particular ground-based transmitter. The typical GPS receiver's location outputs, therefore, cannot be directly used for navigation in relation to a fixed point, as with a VOR or NDB. Rather, an area navigation scheme with waypoint information is used—a simple task in this age of programmable memory computers.

Although there is a technical distinction between positioning and navigation, the two terms are employed interchangeably with the minor discord of difference generally posing no major problems for GPS users. Just remember, the Global Positioning System can be used for navigation purposes.

Ranging

GPS position determination is based on a concept called TOA (time of arrival) ranging, which is a complex way of saying signal travel time from one point to another.

A simple example of TOA ranging is the sending (transmission) of a signal at a precise and known time and the arrival (reception) of the same signal at a later precise and known time. The interval between the time of transmission and the time of reception is the TOA value.

Examples of TOA ranging

As a slow-speed example of the TOA ranging concept, envision an electrical (thunder) storm. The distance from a storm to your location can be simply calculated by counting the seconds from when the visual flash is seen until the thunder is heard. There is a distance of approximately one mile for each five seconds of delay between the flash and the report of thunder (Fig. 2-1). Simply multiply the

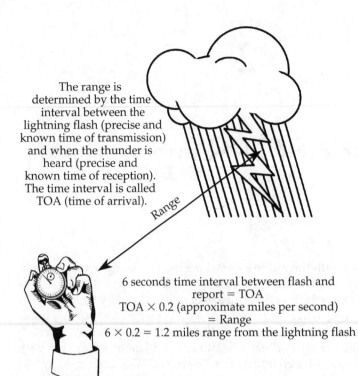

The range is determined by the time interval between the lightning flash (precise and known time of transmission) and when the thunder is heard (precise and known time of reception). The time interval is called TOA (time of arrival).

Range

Fig. 2-1. Nearly everyone is familiar with counting one-one thousand, two-one thousand, three-one thousand, etc. to determine how far away a lightning flash was. This practice is really a primitive form of time of arrival (TOA) ranging.

6 seconds time interval between flash and report = TOA
TOA × 0.2 (approximate miles per second) = Range
6 × 0.2 = 1.2 miles range from the lightning flash

TOA value (time between flash and sound) by 0.2 (speed of sound is approximately two-tenths of a mile per second) to determine the range between you—the receiver—and the storm—the transmitter.

As a further example of the TOA ranging concept, assume the transmitter is a foghorn that blows exactly on the minute mark—precise and known time of transmission—and the receiver is a mariner with a chronometer—accurate timepiece—to be used for timing purposes.

In this example, the foghorn blows exactly on the minute—precise and known time of transmission. The foghorn's sound arrives at the mariner's position exactly 10 seconds after the minute mark—precise and known time of reception. The mariner then multiplies the TOA value of 10 seconds by the speed of sound (1088 feet per second at sea level at 32 degrees F), resulting in a range of 10,880 feet.

If the same mariner could calculate his range from a second foghorn, a position relative to the two foghorns could be determined. Merely draw a circle (TOA-based range circle), centered on the known foghorn positions, with the radius equal to the determined range, for each foghorn. Where the circles intersect is the mariner's location (Fig. 2-2A).

This simple TOA ranging concept works quite well when two conditions are satisfied:

1. That the mariner begins with a good estimate of general position and,
2. That the mariner's chronometer is correct.

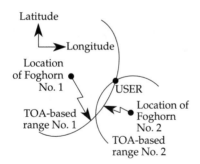

A. TOA-based position determination

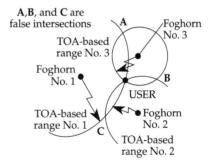

B. Eliminating false intersection ambiguity

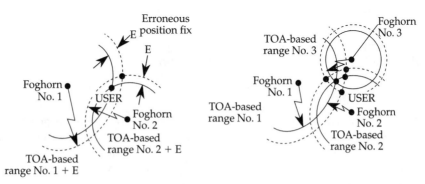

C. Effect of a chronometer bias

D. TOA-based position and time determination

Fig. 2-2. A through D depicts marine TOA-based determinations using a foghorn.
U.S. Government Publication YEE-82-009D

Note: There are two points of intersection between the two range circles. This is why the mariner must have a good idea of present location, otherwise the incorrect intersection point might be chosen.

To reduce intersection ambiguity, range measurements can be made to three foghorns of known location. This will increase the number of intersection points, however, only one intersection point will be common to all the TOA-based range circles; the other intersections will be false (Fig. 2-2B).

TOA ranging and time accuracy

The chronometer used for time measurement during TOA ranging must be very accurate, or precision positioning will not be possible. An error in clock accuracy (called a time-bias error) is, however, correctable.

The effect of time-bias error is erroneous intersection points (Fig. 2-2C). If the mariner's chronometer was running one second fast, then the sound of the foghorn that arrived 10 seconds past the minute mark would appear to have arrived 11 seconds past the minute mark. This error would cause the mariner to compute an erroneous range of 11,968 feet from that foghorn—a 1088 foot error. As the same chronometer would be used to note the TOA from other foghorns, all the TOA-based range observations would have range errors of 1088 feet (shown as E in the example). If range observations were made from only two foghorns, the mariner would have an erroneous position fix. However, the problem of time-bias errors and ambiguous intersections can be solved.

Using the same means as before, note there are three dual-foghorn intersections near the mariner's true position (Fig. 2-2D). Distance E between the intersections of range circles from foghorns one and two, foghorns one and three, and foghorns two and three is strictly a function of the chronometer's time bias. By adjusting the range measurements forward or backward until the three dual-foghorn intersections converge at the true position, the chronometer's time bias can be zeroed out.

As a result of using three transmitters with certain known facts about each (exact position and time of transmission), the ranging information can be applied to a map for determining latitude and longitude coordinates. Additionally, the time-bias error becomes a known and correctable quantity.

This example was based upon the speed of sound, which travels much slower than the speed of light (propagation speed of radio signals used by GPS).

GPS ranging

Orbiting NAVSTAR satellites are the broadcast beacons (transmitters) at the center of TOA-based three-dimensional range spheres (not the two-dimensional circles used for surface positioning). Their signals are sent at the speed of light (186,000 miles per second) and consist of PRN (pseudorandom noise) modulated L-band radio waves. The PRN sequences, C/A (coarse/acquisition)-codes, and P (precision)-codes are predetermined strings of one and zero data bits generated by an on-board clock that also provides the exact transmit time of the broadcasted signals (precise and known time of transmission).

The GPS satellites transmit radio signals via spread spectrum techniques on two frequencies (Fig. 2-3), known as L1 and L2:

L1 = 1575.42 MHz
L2 = 1227.6 MHz

GPS satellite transmitters are digitally modulated with BPSK (bi-phase shift keying) of the carrier. PM (phase modulation) radio differs from AM (amplitude modulation) and FM (frequency modulation) in that changes in the carrier-wave phase are used, rather than changes in the carrier-wave amplitude or frequency, to carry information (Fig. 2-4). BPSK reverses the carrier phase when the digital PRN codes, which are long sequences of ones and zeros appearing to be random in fashion, change from 0 to 1 or from 1 to 0. Additionally, the GPS phase modulated radio system differs from AM/FM radio in that all the satellites transmit on the same frequency. Individual satellites are identified and separated by use of their unique transmitted code sequences.

Although suppressed, technically the carrier bandwidth is 20 MHz for the P-code and 2 MHz for the C/A-code. Phase shift of the carrier causes a spreading of

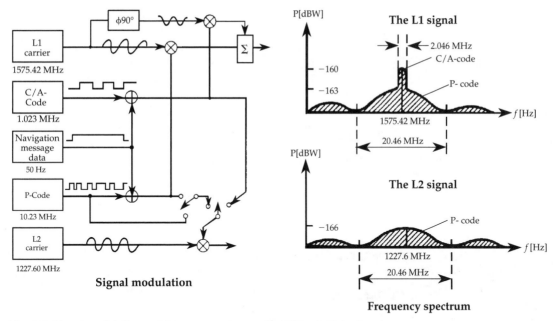

Signal modulation

The L1 signal

P[dBW]

−160

−163

2.046 MHz
C/A-code

P- code

1575.42 MHz

20.46 MHz

f [Hz]

P[dBW]

The L2 signal

−166

P- code

1227.6 MHz

20.46 MHz

f [Hz]

Frequency spectrum

Fig. 2-3. Signal modulation and frequency scheme of a GPS satellite's signal. U.S. Government Publication YEE-82-009Dx

carrier power ±10.23 MHz (from the center frequency) for P-code BPSK and ±1.023 MHz for C/A-code BPSK. The resulting waveform is equivalent to a carrier spread by a regular square wave function at P-code and C/A-code modulation rates.

The P- and C/A-codes are predictable relative to the start time of the code sequence. Therefore, the user can precisely replicate the same code the satellite will send (computed by the GPS-user's receiver/processor).

By mixing the replica P- or C/A-code sequence with the incoming L-band radio signals (from the NAVSTAR satellite) and skewing (offsetting) the replica sequence forward or backward in time, a correlation (matching of the incoming signal with the replica) will be made. The amount of offset made by the receiver's code generator to make the correlation is directly proportional to the range between the receiver and the satellite, and is called the observed TOA value. If the receiver continues to track the signal, while maintaining the correlation, the TOA values may be sampled whenever required (Fig. 2-5).

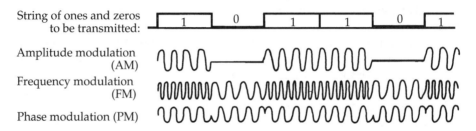

String of ones and zeros to be transmitted:

1 0 1 1 0 1

Amplitude modulation (AM)

Frequency modulation (FM)

Phase modulation (PM)

Fig. 2-4. AM, FM, and PM modulation examples of the sample string of ones and zeros shown at the top. U.S. Government Publication YEE-82-009D

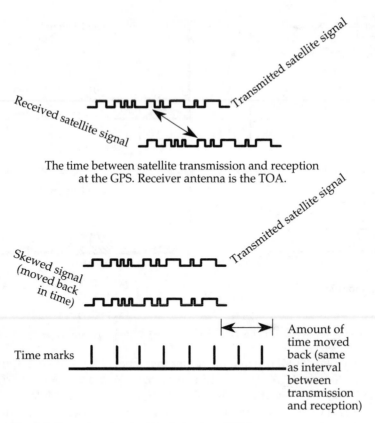

The time between satellite transmission and reception at the GPS. Receiver antenna is the TOA.

Fig. 2-5. Examples showing the derivation of TOA.

Clock bias

If the GPS receiver's clock was synchronized exactly to the on-board satellite clocks, the TOA values observed by the receiver would be equal to the actual geometric ranges between the satellites and the user divided by the speed of light (similar to the mariner, after the chronometer time bias has been zeroed and synchronized with the foghorns). However, in GPS it is not practical to adjust the receiver's clock to zero the time bias. Unlike the mariner using signals traveling at the speed of sound with a chronometer needing adjustment only to within a few thousandths of a second, GPS works with radio signals (traveling at the speed of light) and requires clock accuracy to within a few billionths of a second (a billionth of a second error equals about one foot). Specifically, clock stability of the GPS is 10^{-13} or .003 seconds per one thousand years!

To resolve this time problem, the GPS receiver's clock is left free-running, while the data processor in the receiver mathematically determines the amount of adjustment required to zero the clock's time bias. As a result of the processor's computations, the receiver's observed TOA values are the actual range from each satellite divided by the speed of light plus the time-bias adjustment. They are called PR (pseudorange) measurements because they are similar to measuring the range from the satellites except for the range error of the GPS receiver's clock time bias (Fig. 2-6).

The following definition is the heart of GPS: *A pseudorange measurement is equal to the GPS receiver's observed TOA value multiplied by the speed of light, when the observed TOA value includes both the signal propagation delay due to the actual geometric range and the GPS receiver's clock bias.*

Briefly going back to the mariner, note that the receiver's antenna (ear), accurate clock (chronometer), plotting table (charts), and calculator (data processor) are replaced by one unit for GPS. This combined piece of equipment, generally referred to as the receiver/process or just plain GPS receiver, does all the positioning calculations automatically.

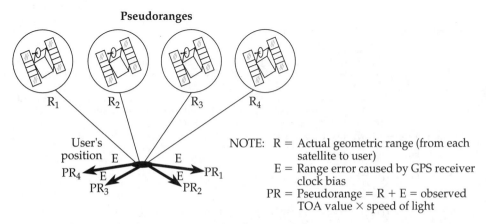

Fig. 2-6. Pseudoranges and what they consist of. U.S. Government Publication YEE-82-009D

Position and time computations

When the GPS receiver begins tracking the PRN sequences from four satellites, and generating TOA values, the receiver's data processor takes over.

By sampling the TOA values from the GPS receiver for each of four satellites, it multiplies them by the speed of light to produce four PR measurements, then compensates the PR measurements for deterministic errors including: the difference between each satellite's clock and GPS system time, the atmospheric distortion of the signals (radio waves travel in straight lines only in a vacuum), effects of relativity, and receiver noise (Fig. 2-7).

The receiver's data processor obtains the information it needs to make these compensations from the navigation message (NAV-msg), which is transmitted from the satellites. Technically, the NAV-msg is superimposed on both the P-code and the C/A-code signals at a data rate of 50 bits/sec and contains 25 data frames, each consisting of 1500 bits divided into subframes of 300 bits each. It takes a receiver 30 seconds to receive one data frame and 12½ minutes to receive all the data frames available (25). Subframes 1, 2, and 3 repeat the same 900 bits of data on all

A. Data processor obtains pseudorange measurements (PR_1, PR_2, PR_3, PR_4) from four satellites

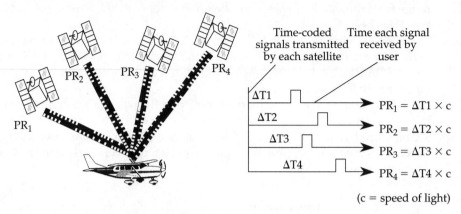

Time-coded signals transmitted by each satellite   Time each signal received by user

$PR_1 = \Delta T1 \times c$

$PR_2 = \Delta T2 \times c$

$PR_3 = \Delta T3 \times c$

$PR_4 = \Delta T4 \times c$

(c = speed of light)

B. Data processor applies deterministic corrections

PR1 = Pseudorange (l = 1, 2, 3, 4)
- Pseudorange includes actual distance between satellite and user plus satellite clock bias, atmospheric distortions, relativity effects, receiver noise, and receiver clock bias
- Satellite clock bias, atmospheric distortions, relativity effects are compensated for by incorporation of deterministic adjustments to pseudoranges prior to inclusion into position/time solution process

C. Data processor performs the position/time solution

Four ranging equations:

$$(X_1 - U_X)^2 + (Y_1 - U_Y)^2 + (Z_1 - U_Z)^2 = (PR_1 - CB \times c)^2$$

$$(X_2 - U_X)^2 + (Y_2 - U_Y)^2 + (Z_2 - U_Z)^2 = (PR_2 - CB \times c)^2$$

$$(X_3 - U_X)^2 + (Y_3 - U_Y)^2 + (Z_3 - U_Z)^2 = (PR_3 - CB \times c)^2$$

$$(X_4 - U_X)^2 + (Y_4 - U_Y)^2 + (Z_4 - U_Z)^2 = (PR_4 - CB \times c)^2$$

X_l, Y_l, Z_l = Satellite position (l = 1, 2, 3, 4)
- Satellite position broadcast in 50 Hz navigation message

Data processor solves for:
- U_X, U_Y, U_Z = User position
- CB = GPS receiver clock bias

Fig. 2-7. Simplified user position/time computation process. Based on U.S.Government Publication YEE-82-009D illustration

25 frames, typically allowing the receiver to capture NAV-msg data within 30 seconds of signal acquisition (Fig. 2-8).

The NAV-msg contains GPS system time of transmission, a hand-over-word (HOW) for the transition from C/A- to P-code tracking, ephemeris (orbital position data), clock data (for the particular satellite being tracked), and almanac data for the remaining satellites in the constellation. Additionally, coefficients for calculating UTC (Universal Time Coordinated) and the ionospheric delay model for

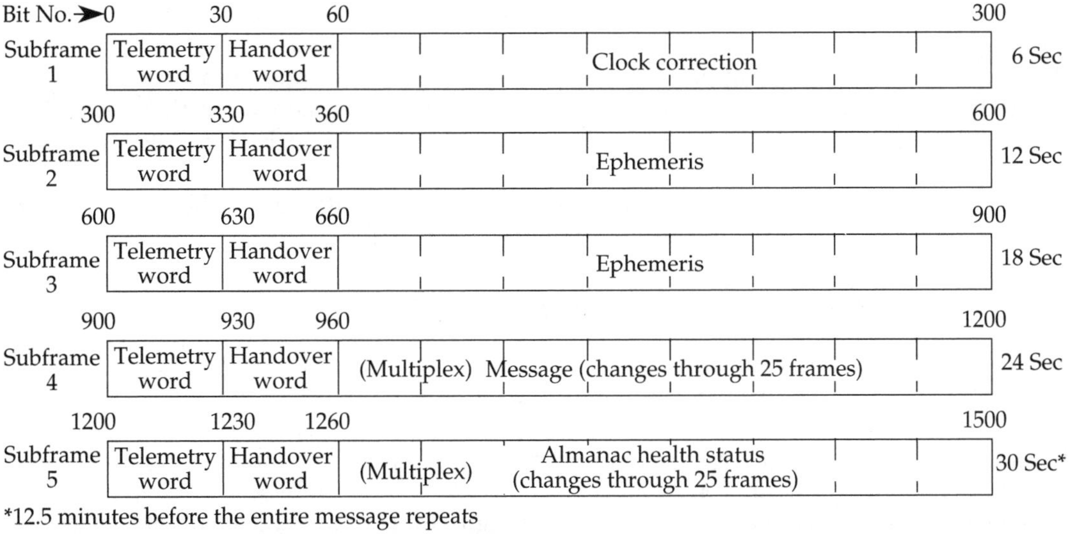

Bit No.➤0		30	60						300	
Subframe 1	Telemetry word	Handover word				Clock correction				6 Sec

	300		330	360					600	
Subframe 2	Telemetry word	Handover word				Ephemeris				12 Sec

	600		630	660					900	
Subframe 3	Telemetry word	Handover word				Ephemeris				18 Sec

	900		930	960					1200	
Subframe 4	Telemetry word	Handover word	(Multiplex) Message (changes through 25 frames)							24 Sec

	1200		1230	1260					1500	
Subframe 5	Telemetry word	Handover word	(Multiplex)	Almanac health status (changes through 25 frames)						30 Sec*

*12.5 minutes before the entire message repeats

Fig. 2-8. Breakdown of the NAV-msg. U.S. Government document

C/A-code users are sent. The data in the NAV-msg is normally valid for a four-hour period.

After making all the necessary adjustments to the PR measurements, the data processor performs the position/time solution calculation to determine its location. The position/time solution process may be thought of as mathematically solving a set of four ranging equations, using the four PR measurements to determine four unknown quantities.

The four unknown quantities are the user's X-position coordinate, Y-position coordinate, Z-position coordinate, plus the time bias (sometimes referred to as the CB or clock bias). This process is not at all unlike the mariner plotting out a two-dimensional position fix and zeroing the chronometer time-bias error using the intersection of three TOA-based range circles. As GPS is a three-dimensional positioning system (X, Y, Z), a fourth TOA-based range sphere is needed. Movement of the GPS satellites is of no consequence, for the NAV-msg contains the information required by the data processor to compute the satellite's exact position at any point in time.

The receiver/processor computes the X, Y, Z position fix in coordinate terms. For user purposes this means an output of latitude, longitude, and altitude.

Note: A GPS position is referenced to the electrical phase center of the receiver's *antenna* (emphasis added), not to the location of the receiver/processor. The antenna is the actual point of signal reception, therefore the antenna's position is determined.

The receiver's data processor can precisely transfer UTC-referenced time to users. The PPS (precise positioning service used by the military) can determine and transfer UTC with an accuracy of 100 nanoseconds and the SPS (standard positioning service available to all others) with an accuracy of approximately 170 nanoseconds.

Under certain conditions, a receiver/processor can compute a position/time fix using PR measurements from fewer than four satellites. This action requires the data processor to receive external-aiding information such as altitude measurements from an accurate barometric altimeter indicating exact antenna location above MSL (mean sea level) or the precise and exact GPS system time. Each aiding source can replace one satellite-based PR measurement in the solution process. For example, if both altitude and time aiding were available it would be possible to solve for PVT using PR measurements from only two satellites.

SIGNAL ACQUISITION AND USE

When the GPS receiver detects the satellite's spread-spectrum broadcast, the signal level is below the earth's natural radio noise level. After detection, the satellite's signal is multiplied by use of the receiver-predicted P- and C/A-codes and the signal collapsed into the original carrier frequency band, concentrated, and brought well above the natural noise level.

The minimum signal power levels (signal strength) for the different signals at a GPS receiver antenna are:

L1 C/A–160 dBW
L1 P–163 dBW
L2 P–166 dBW

A typical satellite tracking sequence begins with the receiver determining which satellites are visible for tracking. Satellite visibility is based on the user-entered predictions of present PVT and on the receiver's stored satellite-almanac data. If almanac data does not exist in the particular receiver, a search of the sky must be made to locate and lock onto any satellite in view. Once a satellite is located and locked onto, the receiver can strip the coded message and read the NAV-msg to obtain current almanac information about the other constellation satellites. Although sounding complicated, the user can relax, as these functions are done automatically by the receiver/processor.

Carrier and code tracking

A carrier-tracking loop is used to track the carrier frequency while a code-tracking loop is used to track the C/A- and P-code signals. The two tracking loops work together in an interactive process, aiding each other, to acquire and track satellite signals.

The receiver's carrier-tracking loop locally generates an L1 carrier frequency which differs from the received carrier signal due to an offset of the carrier frequency (Doppler effect). The Doppler offset is proportional to the relative velocity along the line of sight between the satellite and the receiver antenna plus a bias residual in the receiver's frequency standard. The satellite signal is code correlated (skewed) to allow the carrier signal to become visible and let the carrier-tracking loop track the incoming signal.

The code-tracking loop adjusts the frequency of the receiver-generated carrier until it matches the incoming carrier frequency, thereby determining the relative velocity between the GPS receiving antenna and the satellite being tracked.

The code-tracking loop is used to make the pseudorange measurements between the GPS satellites and the receiver/processor's antenna. The center frequency of the code-tracking loop generated replica of the targeted satellite's C/A-code replica is set by using the Doppler-estimated output of the carrier tracking loop.

The GPS receiver/processor uses the relative velocity of the four satellites being tracked to determine the velocity of the receiving antenna and, for aviation, the airplane it is attached to. The velocity output of the carrier tracking loop is also used to aid the code tracking loop.

P-Code signal acquisition

The C/A-code repeats every millisecond allowing for a minimal receiver search window, however, the P-code repeats only every seven days. This requires that the approximate P-code phase be known in order to obtain signal lock-on. The HOW (hand-over-word) contained in the NAV-msg gives this P-code phase information.

P-code receivers use the HOW from the NAV-msg and the C/A-code derived pseudorange to minimize the P-code search-window requirements. Alternately, a P-code receiver can attempt to acquire the P-code directly, without first acquiring the C/A-code. A drawback with this method is the requirement that the receiver/processor position and exact GPS time be known. The latter generally requires an external atomic clock (due to accuracy needs).

Degraded service

During periods of high jamming—not expected to be a problem for general aviation, marine, or other civilian use—the receiver/processor may not be able to maintain both code and carrier tracking.

This should not present a great technical obstacle for GPS, as the receiver/processor has the capability to maintain code tracking even when carrier tracking is no longer possible. When only code tracking is available, the receiver/processor will skew the locally-generated carrier and code signals based on predicted rather than measured shifts. These predictions (replications) are performed by the processor, which may have additional information available (i.e., almanac data, accurate time aiding, or PVT aiding).

SERVICE AND ACCURACY OF GPS

In navigation, the accuracy of an estimated or measured position of a craft (vehicle, aircraft, or vessel), at a given time, is the proximity of that position to the true position of the craft at that time. The service GPS is able to provide to users is a sum of many factors:

Availability. The availability of a navigation system is the percentage of time that the services of the system are usable. It is an indication of the ability of the system to provide usable service within the specified coverage area.

GPS will provide availability approaching 100 percent, to be further refined based on orbital experience. This figure is based upon a 24 satellite constellation with at least four satellites in view above a five degree masking angle (Fig. 2-9). The 24-satellite constellation is designed to provide worldwide three-dimensional coverage.

National Emergencies. The availability of accurate navigation signals at all times is essential for safe navigation. Conversely, guaranteed availability of optimum performance may diminish national security objectives, therefore contingency security planning is necessary. The United States national policy is that all radionavigation signals (LORAN-C, Omega, VOR/DME, TACAN, GPS, Differential GPS, Transit, and radiobeacons) will be available at all times except during a dire national emergency as declared by the NCA (national command authority), when only those radionavigation signals serving the national interest will be available.

Coverage. The coverage provided by a radionavigation system is that area in which the signals are adequate to permit accurate determination of position. Factors influencing coverage include: system geometry, signal power levels, receiver sensitivity, atmospheric noise conditions, and others which affect signal availability.

Reliability. The reliability of a navigation system is the probability that a system will perform its designed function for a specified period of time under given operating conditions. For example, GPS Block II satellites have a design life of 7.5 years.

GPS reliability figures can be determined only after the satellites are launched and data are collected and evaluated. With the planned replenishment strategy, a constellation of 24 satellites will provide a 98 percent probability of having 21 or more satellites operational at any time.

System Capacity. System capacity is the number of users that a system can accommodate simultaneously. In the case of GPS, the number is virtually limitless due to the system's passive type operation.

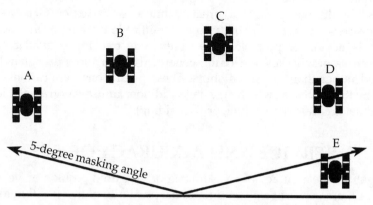

Satellite E is below the 5-degree mask, therefore not usable

Fig. 2-9. Availability of a satellite requires it be located greater than five degrees above the horizon. Satellites A, B, C, and D are available. Satellite E is not available.

Accuracy

The following is quoted from the FRP (Federal Radionavigation Plan) and serves to clarify what some of the many accuracy terms used for GPS mean:

Statistical measure of navigation system errors generally follow a known error distribution. Therefore, the uncertainty in position can be expressed as the probability that the error will not exceed a certain amount. A thorough treatment of errors is complicated by the fact that the total error is comprised of errors caused by instability of the transmitted signal, effects of weather and other physical changes in the propagation medium, errors in the receiving equipment, and errors introduced by the human navigator. In specifying or describing the accuracy of a system, the human errors usually are excluded. Further complications arise because some navigation systems are linear, while others provide two or three dimensions of position.

When specifying linear (one-dimensional) accuracy, or when it is necessary to specify requirements in terms of along-track or cross-track, the 95 percent confidence level will be used. Vertical or bearing accuracies will be specified in one-dimensional terms (2 sigma), 95 percent confidence level.

When two-dimensional accuracies are used, the 2 drms (distance root mean squared) uncertainty estimate will be used. Two drms is twice the radial error drms. The radial error is defined as the root-mean-square value of the distances from the true location point of the position fixes in a collection of measurements. The GPS 2 drms accuracy will be at 95 percent probability.

DOD specifies horizontal accuracy in terms of Circular Error Probable (CEP—the radius of a circle containing 50 percent of all possible fixes). Note that for FRP purposes, the conversion of CEP to 2 drms has been accomplished by using 2.5 as the multiplier (Fig. 2-10).

Types of accuracy

Specifications of radionavigation system accuracy generally refer to one or more of the following definitions:

Predictable accuracy—The accuracy of a radionavigation system's position solution with respect to the charted solution.

Repeatable accuracy—The accuracy with which a user can return to a position whose coordinates have been measured at a previous time with the same navigation system.

Relative accuracy—The accuracy with which a user can measure position relative to that of another user of the same navigation system at the same time.

Accuracy development

GPS was developed to support the broadest possible spectrum of users with accuracy levels established to satisfy their various needs. There are two basic categories of GPS services:

1st ring (inner) is CEP at 50% of the fixes
2nd ring is RMS at 63% of the fixes
3rd ring is 2DRMS at 95% of the fixes
4th ring (outer) is 3DRMS at 100% of the fixes

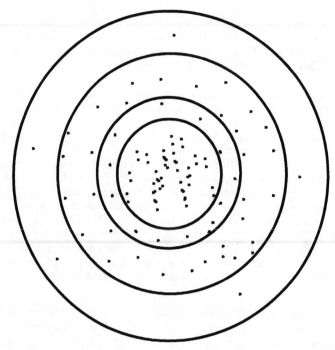

CEP (circle error probability) is based upon a circle of one half the fixes of a specified point. For example: 100 fixes, the CEP would encompass 50.

rms (route mean square) includes 63 of the original 100 fixes.

2drms (twice the distance root mean square) includes 95 of the fixes (95%).

3drms (three times the distance root mean square) includes 100 fixes (100%).

Fig. 2-10. The rings of accuracy describe CEP and the RMS figures used for the definition of GPS accuracy.

PPS (precise positioning service)—A highly accurate positioning, velocity and timing service which is made available only to authorized users (generally this means the military).

SPS (standard positioning service)—A less accurate positioning and timing service which is available to all GPS users, including civilian aviation.

Functionally, the PPS and SPS are nearly identical. The essential difference between them is the attainable accuracy.

Precise positioning service

The precise positioning service is specified to provide 16 meter (52.49 feet) Spherical Error Probable (SEP) positioning accuracy, 100 nanosecond UTC time transfer accuracy, and 0.1 meters (.3937 inches) per second velocity accuracy (depending on the specific receiver design). P-code capable military user equipment will provide a predictable positioning accuracy of at least 22 meters (72.18 feet) (2 drms) horizontally and 27.7 meters (90.88 feet)(2 sigma) vertically and timing/time interval accuracy within 90 nanoseconds (95 percent probability).

PPS is intended primarily for military purposes and authorization to use PPS is determined by the U.S. Department of Defense (DOD), based on U.S. defense requirements and international commitments. Typical authorized users include U.S. military, NATO military, and other selected military and civilian users such as the Australian Defense Forces and the DMA (Defense Mapping Agency).

PPS access is controlled by two features using cryptographic techniques:

1. The SA (selective availability) feature is used to reduce the GPS position, velocity, and time accuracy by inserting controlled errors into the satellite signals.

   Note: The DOD states that during peacetime the effects of SA will be minimized to provide 100 meters (328.1 feet) horizontal accuracy for SPS users, however, system accuracy can be further degraded if the necessity arises.

2. An A-S (anti-spoofing) feature is invoked at random times, without warning, to prevent potential spoofing (hostile imitation) of PPS signals. A-S alters the P-code cryptographically into a code denoted as the Y-code (the C/A-code remains unaffected).

Encryption keys and techniques are provided to PPS users which allow them to remove the effects of SA and A-S and thereby attain the maximum available accuracy of GPS. PPS capable receivers that do not have the proper encryption keys installed will suffer from accuracy degradations and be unable to track the Y-code.

PPS receivers can use either the P(Y)-code or C/A-code or both. Maximum accuracy is attained when using the P(Y)-code on L1 and L2 frequencies. The difference in propagation delay between the two frequencies is used to calculate ionospheric corrections (time delays). C/A-only receivers use an ionospheric model to calculate ionospheric corrections as the C/A-code is broadcast only on L1, making dual frequency delay measurements impossible. This will result in somewhat less positioning accuracy than the dual-frequency P(Y)-code receivers.

Standard positioning service

The standard positioning service is specified to provide 100 meter (328.1 feet) horizontal positioning accuracy to any GPS user during peacetime and attain 170 nanosecond UTC time transfer accuracy. The SPS horizontal accuracy specification

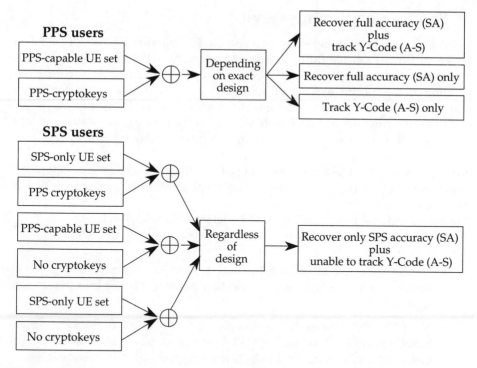

Fig. 2-11. PPS and SPS GPS receiver capabilities/limitations. U.S. Government Publication YEE-82-009D

includes peacetime degradation of SA (the dominant SPS error source) and the ionospheric modeling error (Fig. 2-11).

SPS is intended primarily for civilian purposes, although it has many peacetime military uses, and is the service that general aviation will use.

Accuracy of position and time

When GPS is declared operational, the DOD plans to provide, on a daily basis at any position worldwide, horizontal-positioning accuracy within 100 meters (328.1 feet) (2 drms) and 300 meters (984.2 feet) with 99.99 percent probability. In practice, much greater accuracy is the rule and not the exception.

The accuracy of receiver/processor position and time solutions are determined by two very important factors:

1. Error in the measurement of pseudoranges from each satellite being tracked and
2. Satellite-to-user geometry

Knowing these factors for a specific GPS receiver at a particular point in time and space is important as it allows an understanding of the limitations of GPS and enables forecasting of the receiver's position and time accuracy.

Accuracy limiting factors

The error in the receiver's measurement of the PRs from each satellite is called the user equivalent range error (UERE) and is the product of:

- Stability of the particular satellite's clock
- Predictability of the satellite's orbit
- Errors in the satellite-broadcast NAV-msgs
- Precision of the GPS receiver PRN sequence tracking design
- Errors in the processor's calculation of the ionospheric model

The UERE depends, in part, on the quality of the broadcasted satellite signals and will vary between satellites and from time to time. It also depends on receiver/processor designs and can therefore vary from equipment to equipment (Fig. 2-12).

The second accuracy-limiting factor is called the dilution of precision (DOP), which is a geometric quantity depending upon the relative positions of the user and the selected satellites. High values of DOP cause small range measurement errors to become large position errors. Therefore, the four satellites used by the receiver/processor to determine its PVT should be selected to minimize high-DOP values. The limiting factor is independent of the quality of the broadcast satellite signals or the type of GPS receiver (providing the same four satellites are selected). DOP is, basically, an amplification factor that multiplies the UEREs and increases the receiver/processor's units position/time solution errors.

A good DOP means the satellites exhibit good geometry as seen by the receiver/processor. Good DOPs are low numbers while poor DOPs are high numbers. Low numbers occur when the satellites are widely spaced in the sky above the GPS receiver (Fig. 2-13).

High DOPs occur when the satellites are close together or when they form a row or circle. Although rare, it is possible for DOP factors to be so large they prevent the receiver/processor from processing a solution (Fig. 2-14).

User equivalent range error (UERE):

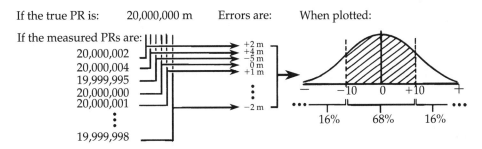

Fig. 2-12. UERE depends upon satellite signal quality and receiver/processor equipment design.
U.S. Government Publication YEE-82-009D

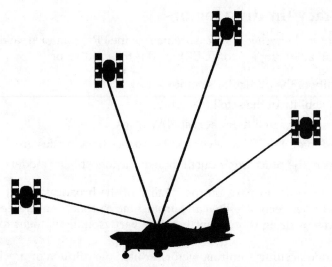

Fig. 2-13. Good DOP requires that the four selected satellites be widely spaced above the user.

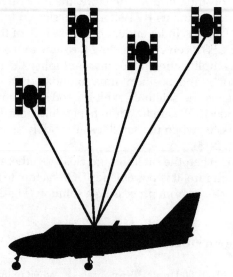

Fig. 2-14. Poor DOP results from the selected satellites being too close together.

Note: The best possible geometry for a receiver/processor is for one satellite to be directly overhead and three other satellites equally spaced around the horizon.

There are special types of DOPs for each of the position and time solution dimensions:

VDOP (vertical or spherical dimension) describes the effect of satellite geometry on the receiver/processor altitude errors.

HDOP (horizontal or circular dimension) describes the effect of satellite geometry on latitude/longitude errors.

PDOP (position) a combination of HDOP and VDOP.

TDOP (time) in relation to the VDOP, HDOP, PDOP.

GDOP (geometric) relates to satellite positions relative to the user.

Position and time errors

Knowing the UERE and DOP (values seen by a specific receiver/processor at a particular time and place) allows the forecast of that unit's position-time errors. The ability to forecast is essential for real-time performance estimation within the receiver/processor and, planning and analyzing the use of GPS.

Navigation and time error forecasting is easy when all satellite signals give the same UERE. In such a case, the forecast navigation and time errors can be computed by simply multiplying the appropriate DOP value by the common satellite UERE value.

DOPs can be forecast as a function of the user's location and time, as well as the number of satellites visible and their locations. The receiver/processor's portion of the UERE is usually calibrated (compensated for) by the manufacturer. The satellite portion of the UERE is forecast by the control segment and provided to the users as a part of the NAV-msg.

Accuracy projections and simulations

Accuracy projections for GPS are based upon computer simulations. At a specified time of day, the positions of the GPS satellites are calculated, by the master control station at Falcon AFB, to determine which are visible at a given location on earth. Four of the visible satellites are selected (for that location) and the location solution that a GPS receiver would provide is calculated. As a receiver/processor determines location by estimating the range to each of the four satellites, the simulations mimic real errors in the process by introducing a range error for each of the simulated satellites, using mathematical techniques. The range data are used to solve for the user's location and the instantaneous position error is determined by subtracting the true position from the calculated position.

By repeating this process for many locations around the earth over a 24-hour period, the simulations produce a composite view of system performance. These results are dependent upon several program inputs:

- The number of satellites in the GPS constellation
- The orbits chosen for the satellites
- The locations of the simulated users
- The local visibility constraints on receiving signals from satellites
- The criteria for selecting four satellites from among those visible

All accuracy projections are based upon a fully-operational system with 24 healthy satellites and normal uploads by the control segment. Satellite visibility depends upon local conditions. Some users may be able to track satellites less than 5 degrees above the horizon, while other users may have difficulty even at 10 degrees. DOD accuracy simulations use 5 degrees.

Accuracy simulations use the four-satellite combination that minimizes three-dimensional position DOP (PDOP). In some applications a receiver/processor may have access to additional information (aiding) such as a known altitude or have an accurate atomic clock. In general, such information improves position accuracy substantially.

When discussing horizontal accuracy it is important to differentiate between a user whose horizontal errors are based upon the use of four satellites that minimize DOP, and one based upon a known altitude and the use of three satellites that minimize horizontal DOP (HDOP).

System accuracy specifications

There are four formal specifications for GPS accuracy (Fig. 2-15):

1. PPS user 3-D (spherical) position accuracy shall be 16 meters (52.49 feet) SEP (spherical error probable) or better.
2. PPS user velocity accuracy in any dimension shall be 0.1 meters/second (3.937 inches/second) or better.
3. PPS user time accuracy (with respect to UTC) shall be 100 nanoseconds one sigma or better.
4. SPS user 2-D (horizontal) position accuracy shall be 100 meters (328.1 feet) 2 drms (distance root mean square) or better.

Note: The accuracy of the position and time solutions obtainable from a GPS receiver are the product of two factors:

1. User equivalent range error (UERE) value
2. The dilution of precision (DOP) value

The average UERE and DOP values for the CGSS (composite global sample space) upon which the four GPS accuracy specifications are based on:

UERE—The UERE value for PPS receiver/processors is 7.0 meters (22.96 feet) one sigma while the UERE value for SPS receiver/processors is approximately 32 meters (104.99 feet) one sigma.

DOPs—For PPS receiver/processors the values are: HDOP 1.5, VDOP 2.0, PDOP 2.5, and TDOP 1.1. SPS values are: HDOP 1.6, VDOP 2.2, PDOP 2.7, and TDOP 1.3.

System survivability

The survivability of the GPS is tantamount to its acceptance, integrity, and continued use (Fig. 2-16). To afford survivability the system consists of a satellite constellation of 21 SVs (space vehicles) and 3 in-orbit spares that can be repositioned easily and quickly.

PPS / SPS	50th percentile	drms	2 drms
Position			
• Horizontal	8 m / 40 m	10.5 m / 50 m	21 m / 100 m
• Vertical	9 m / 47 m	14 m / 70 m	28 m / 140 m
• Spherical	16 m / 76 m	18 m / 86 m	36 m / 172 m
Velocity			
• Any axis	0.07 m/sec	0.1 m/sec	0.2 m/sec
Time			
• GPS	17 nsec / 95 nsec	26 nsec / 140 nsec	52 nsec / 280 nsec
• UTC	68 nsec / 115 nsec	100 nsec / 170 nsec	200 nsec / 340 nsec

NOTES:
- Formal GPS system accuracy specifications are shown in the shaded areas.
- Derived GPS system accuracy values are shown in the unshaded areas.
- There is no SPS user velocity accuracy specified.
- 50th percentile is equivalent to CEP, SEP, etc.

Fig. 2-15. Statistical specification chart of GPS accuracy. U.S. Government Publication YEE-82-009

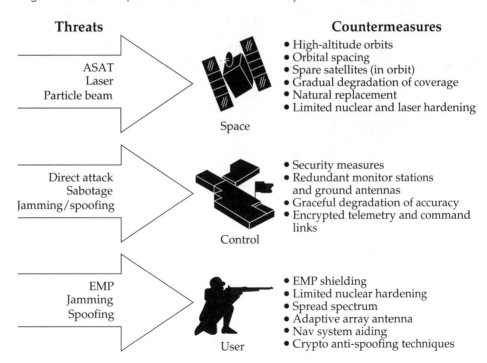

Threats   Countermeasures

Threats

ASAT
Laser
Particle beam

Space

Countermeasures

- High-altitude orbits
- Orbital spacing
- Spare satellites (in orbit)
- Gradual degradation of coverage
- Natural replacement
- Limited nuclear and laser hardening

Direct attack
Sabotage
Jamming/spoofing

Control

- Security measures
- Redundant monitor stations and ground antennas
- Graceful degradation of accuracy
- Encrypted telemetry and command links

EMP
Jamming
Spoofing

User

- EMP shielding
- Limited nuclear hardening
- Spread spectrum
- Adaptive array antenna
- Nav system aiding
- Crypto anti-spoofing techniques

Fig. 2-16. Threats to the survivability of the three segments of GPS and planned countermeasures. U.S. Government Publication YEE-82-009D

Should there be a loss of any GPS satellite(s), replenishment from the ground can be achieved within two months, by launch.

The GPS system is designed to degrade slowly in the event of a loss of satellites. For example: In the event of the loss of 9 SVs, a full 12 hours of 3D and 20 hours of 2D continued daily capability worldwide is expected. Further, due to new designs of the Block II SVs, a loss of communications between the master control station and the satellites would not cause an immediate failure of the system, merely a graceful degradation (Fig. 2-17).

The GPS satellites are designed with a limited amount of laser hardening. Further, it is not currently feasible for ground-based lasers to be effective at the GPS orbital height. The satellites are protected against electromagnetic pulse (EMP) damage.

A nuclear detonation in space would cause a blackout effect lasting 10 or more minutes at L-band frequencies and could incapacitate erasable read-only memories (EROMS) and random-access memories (RAMS) if nearby, however, the capability for restoring erased memories exists.

Jamming of the control system between outstations and the master control station is possible, however, four stations would need to be jammed simultaneously. A hostile takeover of the tracking, telemetry and control aspects of any space vehicle is always a threat, but the uplinks use encrypted signals, thereby giving added protection.

To prevent spoofing from an enemy source, the PPS is further protected by encrypting the P-code. Additionally, many GPS navigation receiver/processors are coupled or integrated with inertial, LORAN, or other navigation systems.

The next generation of GPS satellites (Block IIR) will be equipped for cross-link ranging, making offshore monitoring stations redundant.

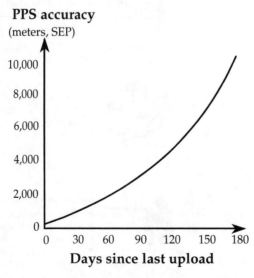

Fig. 2-17. Graceful degradation chart showing the decreasing accuracy as the days from the most recent upload increase.

DIFFERENTIAL GPS

As already mentioned, GPS can exhibit variations of accuracy. These variations may be caused by propagation anomalies, errors in geodesy, accidental disturbances of signal timing, or other factors—specifically SA (the intentional system dithering by the DOD for national security reasons). The adverse effects of these variations may be substantially reduced or eliminated by differential techniques.

In differential operation, a special facility is precisely located at a fixed point within an area of interest. Received GPS signals are observed in real time and compared with the signals expected to be observed at that fixed point. The differences between the observed signals and the predicted signals are transmitted to users as differential corrections to increase the precision and performance of those users' GPS receivers.

How DGPS works

A DGPS reference station is fixed at a geodetically surveyed position. From this position, the reference station tracks all satellites in view, downloads ephemeris data from them, and computes corrections based on its measurements and precise geodetic position. These corrections are then broadcast to GPS users to improve their navigation solution. There are two well-developed methods of handling this:

1. Computing and transmitting a position correction in X-Y-Z coordinates, which is then applied to the user's GPS solution for a more accurate position.
2. Computing pseudorange corrections for each satellite, which are then broadcast to the user and applied to the user's pseudorange measurements before the GPS position is calculated by the receiver, resulting in a highly accurate navigation solution.

The first method, in which the correction terms for the X-Y-Z coordinates are broadcast, requires less data in the broadcast than the second method; but the validity of those correction terms decreases rapidly as the distance from the reference station to the user increases. Both the reference station and the user receiver must use the same set of satellites for the corrections to be valid, a feat often difficult to achieve.

Using the second method, an all-in-view receiver at the reference site receives signals from all visible satellites and measures the pseudorange to each. Since the satellite signal contains information on the precise satellite orbits and the reference receiver knows its position, the true range to each satellite can be calculated. By comparing the calculated range and the measured pseudorange, a correction can be determined for each pseudorange measurement at each user's location. This method provides the best navigation solution for the user, is the preferred method, and is the method now employed by the U.S. Coast Guard DGPS Service (Fig. 2-18).

For a civilian user of the standard positioning service (SPS), differential corrections can improve navigational accuracy from 100 meters (328.1 feet) (2 drms) to better than 10 meters (32.8 feet) (2 drms) (Fig. 2-19).

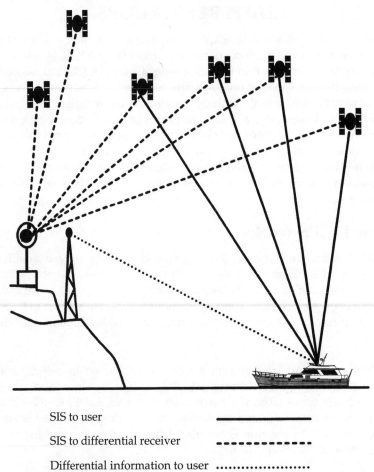

SIS to user _____

SIS to differential receiver - - - - - - - - - - -

Differential information to user ·····················

Fig. 2-18. Differential GPS system currently in use by the USCG transmits correctional data via existing low-frequency beacons for use in harbor navigation. The system was developed by the Radio Technical Commission for Maritime Services (RTCM) for marine use.

Magellan Systems

Fig. 2-19. External DGPS receiver used in conjunction with a GPS receiver.

The military and DGPS

The reason for civilian use of DGPS is to overcome the effects of selective availability (SA) and to provide greater accuracy. Remembering that GPS was designed as a U.S. military tool, enhanced unauthorized (civilian) accuracy is precisely what the military does not wish.

SA was intended to prevent adversaries of the U.S. from having enhanced military capabilities based upon GPS. Unfortunately, DGPS, in particular WADGPS (wide-area differential GPS), is capable of providing accuracies suitable for military needs to anyone, anywhere, at anytime.

It's interesting to note that SA was not used during the Gulf War because of the large number (13,000+) of civilian-type C/A-code GPS receivers in use by our military. This was a prime example of a military situation for which SA was intended, yet it was not used.

If the SA feature was not used, there would be little need for DGPS under most circumstances for civilian use.

Note: The President of the United States may suspend DGPS at any time for reasons of national security.

DGPS coverage

The area over which corrections can be made from a single differential facility depends on a number of factors, including timeliness of correction dissemination, range of the correction transmission, area and uniformity of the system's grid, and user equipment implementations. A differential facility might serve an area with a radius of several hundred miles, depending on the system used and the method of implementation, or be limited to only a few miles. It is possible to develop a DGPS system workable worldwide.

The USCG transmits the DGPS data on LF beacon carriers. The operating frequencies range from 283.5 to 325.0 kHz and can have effective ranges to a couple of hundred miles or more.

Further research and development

The DOD, in coordination with the FAA, is examining differential GPS for use at improvised aircraft landing sites—jungle clearings, interstate highways, etc. The concept is to assemble components that are currently available into an operational system. Every attempt will be made to avoid the requirement for additional aircraft avionics. The objective of this R&D effort is to enhance the benefits of GPS rather than to develop a new precision landing aid.

For general-aviation purposes, the FAA is planning to use differential corrections to GPS in the National Airspace System, including instrument approaches.

3

Civilian aviation use

THE CIVILIAN AVIATION community views GPS quite differently from most military users. The primary technical difference being that the standard positioning service (SPS) C/A-code will be the only GPS signal available to civilian aviation users. Therefore, aviation users must operate with the reduced accuracy caused by the selective availability (SA) limitation.

The primary differences in GPS use between civilian aviation interests and military users are that the military uses GPS for weapons delivery systems and mission navigation, while civil aviation is interested in simple navigation and position reporting. Further, aviation users are interested in assurances of safety-of-life issues as GPS signal failures could potentially affect large areas, and consequently, large numbers of aircraft simultaneously.

GPS equipment for civilian aviation will cover the extremes from minimum-capability stand-alone receiver/processors for general-aviation users to sophisticated GPS/integrated navigation systems for commercial users.

From a governmental point-of-view, GPS has the potential to replace most existing radionavigation aids, thereby reducing the maintenance and upkeep costs of these systems. The primary benefits from the user point-of-view is direct routing and global navigational coverage.

The first large-scale individual use of GPS for navigation of a civilian airplane was in 1990, when Tom and Fran Towle circumnavigated the globe in their Cessna 310 using GPS for much of their guidance. Note that at that time there were only nine satellites available, causing many periods of GPS unavailability.

Aviation regulators and advisors

The responsibility for establishing regulations and certifying the use of GPS in civilian airspace is shared by two regulatory agencies:

ICAO (International Civil Aviation Organization)—a cooperative multinational organization responsible for regulating international air traffic.

FAA (Federal Aviation Administration)—responsible for regulating the NAS (national airspace system).

Two advisory groups charged with developing and recommending standards for the use of GPS in aviation are:

RTCA (Radio Technical Commission for Aeronautics)—to develop a MOPS (minimum operational performance standard) for GPS equipment, which will guide the FAA in adopting appropriate regulations.
EUROCAE (European Organization for Civil Aviation Electronics)—performing a function similar to RTCA in Europe.

Both organizations are voluntary, with no government status, and are composed of representatives from various government, industry, and related private organizations to facilitate broad participation in the GPS standardization process.

The AEEC (Airlines Electronic Engineering Committee) is a cooperative international organization of airline representatives currently developing universal standards for GPS equipment and integrated navigation systems.

Equipment manufacturers and private organizations also participate in committee activities to help minimize acquisition costs, maximize interoperability, and standardize equipment specifications.

Military coordination

Civilian aviation concerns are coordinated with the U.S. military on various levels by the U.S. Department of Transportation and DOD, which jointly developed the FRP (Federal Radionavigation Plan) serving as the planning and policy statement for all U.S. Government radio navigation systems.

The FRP is updated every two years based on a review by DOT and DOD representatives and direct input from the public via a series of radionavigation user conferences.

At the development level, the DOT is a direct participant in the GPS JPO (Joint Program Office) and maintains a Deputy Program Director to represent civilian interests. At the operational level, the Civil GPS Service Steering Committee and the U.S. Coast Guard, via the GPS Information Center, distribute GPS operational information and coordinate civilian user concerns with Space Command and the Control Segment.

Civilian aviation concerns

The primary concerns civilian aviation have about GPS are continued availability, system accuracy, and operational integrity.

As previously mentioned, a GPS signal availability loss or severe accuracy degradation could affect large geographical areas and therefore large numbers of aircraft simultaneously. For this reason, civilian aviation organizations have been strong advocates for maximizing the number of active GPS satellites in order to minimize the effect of losing any particular satellite(s) signal.

System availability and accuracy have been well discussed in previous pages. It is the position of the DOD that neither should present problems to the user.

INTEGRITY

Integrity is defined, in most aviation navigation references, as the ability of a system to provide timely warnings to users when the system should not be used for navigation. Current ground-based radionavigation aids continuously monitor their output signals and shut down when a significant error is detected. Generally, shut down occurs within seconds of detection, which is typically automatic.

Although the control segment and each satellite monitors GPS signal performance, the response time and fault monitoring have not proved sufficient for aviation purposes. Further, the accuracy degradation of SA is statistical in nature and inherently without bounds (from NAVSTAR GPS User Equipment Introduction— NATO).

Some integrity problems are diminished by the control segment's update to the NAV-msg (including a satellite health message which is transmitted by each satellite). The satellite health message is not changed between satellite updates and is transmitted as part of the GPS navigation message for reception by both PPS and SPS users. Further, satellite operating parameters such as navigation-data errors, signal availability/anti-spoof failures, and certain types of satellite-clock failures are monitored internally within the satellite. If such internal failures are detected, users are notified within six seconds via their receiver/processors. However, other failures, which are detectable only by the control segment, may take from 15 minutes to several hours to appear. As pointed out by one professional pilot group, that is a very long time when doing an ILS approach!

DOD concept of integrity

According to DOD's concept of integrity, GPS satellites are monitored more than 95 percent of the time by the network of five monitoring stations. The information collected by the monitoring stations is processed by the master control station and used to update the navigation message.

DOD GPS receivers use the information contained in the navigation and health messages, as well as self-contained satellite geometry algorithms and internal navigation solution convergence monitors, to compute an estimated FOM (figure of merit). This number is displayed continuously to the operator, indicating the estimated overall confidence level of the position information.

DOT and DOD have recognized the requirement for additional integrity for aviation and, as a result, the development of integrity capabilities to meet flight safety requirements is underway.

Integrity requirements and assurances

Depending upon the specific GPS application, supplemental or sole-means of navigation, the requirements for integrity will differ.

In both cases the faults that effect integrity must be detected. However, for sole-means navigation the fault must be isolated so that a defective satellite can be de-selected and navigation allowed to continue.

The two primary approaches to assuring GPS integrity are RAIM (receiver autonomous integrity monitoring) and GIC (GPS integrity channel).

RAIM, as the name implies, consists of analyses that the receiver can perform autonomously or in conjunction with existing on-board navigation aids. A common RAIM technique relies on the principle that the receiver can, in most cases, detect and isolate an integrity failure if it has an over-determined position solution. For example: Five satellite signals are available. Five position solutions can be obtained using combinations of four of the five satellites. If a single satellite exhibits a large pseudorange error, the four solutions based on the faulty satellite will be similar to each other and significantly different from the fifth. In such a case the error can easily be detected and isolated.

The GPS integrity channel consists of a ground-based monitoring system with a satellite broadcast warning and is based on monitoring the GPS satellites from surveyed sites and broadcasting the observed range errors. The GPS receiver then evaluates the errors and selects a combination of satellites that either meets the user's current needs or informs the user that GPS is not presently capable of meeting those needs.

RAIM and GIC can support each other and may be used in combination to meet integrity requirements. Additionally, DOD is investigating means of upgrading the control segment's monitoring and failure response time.

Of note is the fact that GPS has been designed with expandability in mind. Future capabilities will no doubt include ILS (instrument landing system) uses, precise-position reporting for air traffic control and flight following purposes, and aircraft conflict control (collision avoidance).

AVIATION USE

Prior to military FOC (full operational capability), there is expected to be significant civilian use of GPS for marine and air navigation, to obtain accurate PVT data, for geodetic surveying, and for many other applications. Initially, civil aircraft will use GPS as a supplementary system for enroute domestic and international operations and slowly move into some nonprecision operations (i.e., nonprecision instrument approaches).

The current minimum operational performance standard (MOPS) for airborne supplemental navigation equipment using GPS is: At least five satellites in view above a mask angle of 7.5 degrees in which all combinations of four out of five satellites provide the horizontal position accuracy required (Fig. 3-1). At least five satellites are required so that if one satellite fails, unaided GPS navigation may continue.

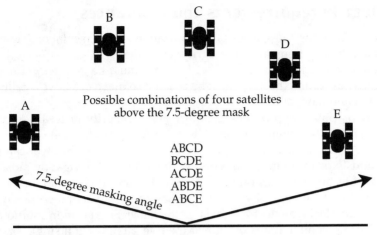

Fig. 3-1. Possible combinations of four satellites above the mask.

The integrity requirement for nonprecision approaches requires a warning to the pilot or removal of the signal from service within 10 seconds after the signal has gone out-of-tolerance.

FRP statement about system longevity

In the Federal Radionavigation Plan the DOD states:

GPS-SPS is planned to be available beginning in 1993 on a continuous, world-wide basis with no direct user charges for a minimum of ten years. The service will provide horizontal accuracies of 100 meters (2 drms—95% probability) and 300 meters (99.99% probability). Beyond the original offer of GPS-SPS for a minimum of ten years, the U.S. intends to continue operation of GPS and to offer GPS-SPS for the foreseeable future free of direct user fees. In addition, the U.S. intends, subject to the availability of funds, to provide a minimum six-year advance notice of termination of GPS operations or elimination of the GPS-SPS.

U.S. military use of the national airspace system

Military use of the national airspace system (NAS) is a cooperative effort between the services and the FAA. Generally, equipment, operational requirements, and flight certification are performed by the individual services, although some military aircraft do maintain civilian flight certification.

Normally, military aircraft do not have the same GPS integrity requirements as civilian aircraft, due to the extensive use of integrated navigation systems and the nullification of SA effects by using PPS.

Availability after IOC

Subsequent to IOC (initial operational capability), any planned disruption of the SPS *in peacetime* (emphasis added) will be subject to a minimum 48-hour advance notice. Such notice will come to aviation users through the FAA's NOTAM (Notice to Airmen) system. A disruption in service is defined as periods in which the GPS is not capable of providing SPS as specified. Unplanned system outages—system malfunctions or unscheduled maintenance—will be announced via NOTAM as they become known.

FRP policy for SPS and PPS

The DOD states very clearly what SPS and PPS are expected to do in the Federal Radionavigation Plan:

> SPS is a positioning and timing service which will be available to all GPS users on a continuous, worldwide basis with no direct charge. SPS will be provided on the GPS L1 frequency which contains a coarse acquisition (C/A) code and a navigation data message. SPS is planned to provide, on a daily basis, the capability to obtain horizontal positioning accuracy within 100 meters (2 drms, 95 percent probability) and 300 meters (99.99 percent probability), vertical positioning accuracy within 140 meters (95 percent probability), and timing accuracy within 340 ns (95 percent probability). The GPS L1 frequency also contains a precision (P) code that is reserved for military use and is not a part of the SPS. Although available during GPS constellation build-up, the P code will be altered without notice and will not be available to users that do not have valid cryptographic keys.
>
> PPS is a highly accurate military positioning, velocity, and timing service which will be available on a continuous, worldwide basis to users authorized by the DOD. PPS will be the data transmitted on GPS L1 and L2 frequencies. PPS was designed primarily for U.S. military use and will be denied to unauthorized users by use of cryptography. PPS will be made available to U.S. Federal and Allied Government users—civilian and military—through special agreements with the DOD. Limited, non-Federal Government, civilian use of PPS, both domestic and foreign, will be considered upon request and authorized on a case-by-case basis.

Military use after IOC

GPS will be integrated into military aircraft that are instrumented for IFR flight and contain inertial navigation systems or other forms of suitable attitude heading reference systems. These aircraft will be flight tested to ensure that they meet established standards for operation in the NAS.

4

GPS applications

WHAT THE DOD ENVISIONED as the mission of GPS was limited to military purposes only. What was desired was accurate and precise positioning, velocity, and time determination on a full-time worldwide basis. Since the first actual deployments of satellites, making the system partially available to potential users, the applications for GPS have been on the increase and not just in the military world. In short: GPS has been discovered!

The tremendous acceptance level of the GPS, for positioning and navigation purposes, comes because of the extreme ease of use, the relatively high-accuracy level, and the excellent reliability factor.

It can be said that many of the applications being made of GPS, and those coming in the future, will directly impact nearly everyone at one time or another.

MILITARY APPLICATIONS

The positioning performance of GPS enhances, and/or will enhance, many areas of military operations. The system's operational simplicity, accuracy, and reliability are the keys that will make it an indispensable military resource (Fig. 4-1).

Military testing has shown substantial improvements in combat applications when supported by the very accurate PPS (precise positioning service). Remember, PPS is more accurate than the SPS (standard positioning service) that civilian user relies upon.

Aviation

The simple facts are that GPS accuracy can streamline enroute, terminal, and approach navigation, resulting in reduced flight times and fuel consumption. As the GPS is a 3D (three-dimensional) system, descent and nonprecision approach and landing can be more closely controlled and, using DGPS (differential GPS), the more position-critical carrier landings and instrument landings will come to de-

Worldwide
24 hours a day

Enroute navigation
Terminal navigation
Low-level navigation
Nonprecision approach
Target acquisition
Close air support
Missile guidance
Command and control
Air drops
Surveying and mapping
Time synchronization
Rendezvous
Bombing
Drone vehicle control
Base/site preparation
Search and rescue
Reconnaissance
Mine placement
Space navigation

Fig. 4-1. GPS military applications.

pend upon GPS. Aircraft rendezvous, such as for in-flight refueling, are made easy using GPS.

In combat-related applications, the PPS performance will improve the accuracy of bombing and ballistic weapon delivery. Close air support can be greatly enhanced and made safer for friendly forces. Reconnaissance and target location becomes as easy as the push of a button to mark the location of each target.

Ground forces

The precise positioning capabilities of GPS will enhance site surveying, field artillery placement, target acquisition and location, and target hand-off operations. Points of observation can be accurately noted for reference or safety reasons.

First-round artillery effectiveness will be improved by precision forward-observer determinations of enemy locations and movements. Additionally, the exact positions of friendly forces will be known, an extremely important point involving all weapons delivery.

In noncombat operations, PPS (and SPS) accuracy will support efficient off-road navigation for supply distribution, equipment recovery, personnel rendezvous, area reconnaissance, etc.

Insertion and extraction missions may be carried out with extreme accuracy, allowing for the safe and timely deployment and evacuation of troops. Further, the same means can provide for easy medevac with little search time required.

Operation Desert Storm used the GPS for land (surface) navigation on the desert areas for many operations, including troop movements and logistical support. It has been rumored that many of the GPS receivers used by the troops were purchased on the open market, some troops spending their own money for the units. The reasoning was simple: quick availability, small size, and SPS accuracy was sufficient for many uses. Maps of the deserts were virtually nonexistent and desert physical/geographical reference points were few and far between.

For the land-based soldier there has never been anything like GPS with its powerful capabilities in a device that fits into the palm of the hand (Figs. 4-2 and 4-3).

Magellan Systems

Fig. 4-2. A hand-held GPS receiver, such as this Magellan NAV 1000M, provides fast, easy position information in an area with few landmarks.

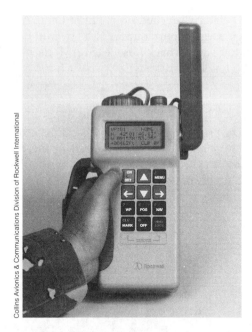

Fig. 4-3. This hand-held GPS unit is designed for the military and uses the highly accurate PPS.

Collins Avionics & Communications Division of Rockwell International

Naval operations

General high-seas navigation becomes very simple using GPS and harbor navigation accuracy can be greatly improved over present methods. Further fine tuning for harbor navigation includes the development and installation of the DGPS.

GPS greatly simplifies the process of coastal surveying and mapping for planned operations—i.e., amphibious operations, coastal patrol, etc. It takes simplicity to the point that only a few buttons need be pushed while over-flying or otherwise traversing the coastline. Later examination of the instant location notations (button pushing) allow chart makers to complete their work.

Placing mines in navigable waters can be accelerated and done with much greater accuracy. Exact placement location maps can later be created for navigation among the explosive devices.

Submarines can position with precise accuracy for updating their inertial guidance systems with a minimal antenna exposure. Actual exposure is only a few inches in diameter and need be barely above the water's surface.

Weapons delivery

GPS was designed to increase the accuracy of military weapons delivery. Using self-contained receivers, missiles can navigate themselves to precisely predetermined targets. The trick is merely locating itself using GPS, knowing where it is going from preprogrammed data, and making the necessary flight adjustments to navigate between the two.

The ship-launched Tomahawk Land Attack Missile (TLAM) uses GPS technology as an important part of its overall navigation system. Similarly, the USN's Standoff Land Attack Missile (SLAM) uses GPS in its cruise missile anti-ship mission (Figs. 4-4 and 4-5).

Time transfers

The precise time-transfer capabilities of GPS will allow global synchronization of all military electronic systems. This means various equipment can be operated in perfect synchronization merely by monitoring the GPS constellation. Among those applications:

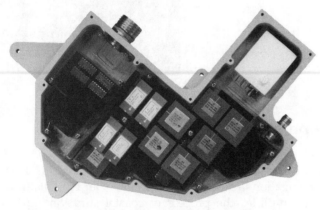

Fig. 4-4. Collins TLAM GPS Receiver designed to guide and position the Tomahawk Missile. Collins Avionics & Communications Division of Rockwell International

Fig. 4-5. Collins SLAM GPS Receiver used in the Standoff Land Attack Missile. Collins Avionics & Communiations Division of Rockwell International

- Secure communications systems using time as an encryption key
- Electronic warfare measures
- Target planning techniques
- Time coordinated military actions

Complex equipment is not necessary at the user site for precise time-transfer. There is no need for the typical "large, heavy, rack-mounted GI-type equipment" generally associated with military operations; the equipment needs to consist only of a receiver/processor. The precision is provided by the GPS and simplicity of field use is the result of careful receiver/processor design.

Interoperability

Because a common grid datum system is used by the GPS, interoperability in the various aspects of air, ground, and sea operations can exist at a level previously unheard of. Most user equipment is capable of output to many different datum systems (via internal conversion software), making it useful with nearly any type of mapping system used. Generally, the readout accuracy of a receiver/processor will exceed the potential accuracy of plotting on a map. Examples of interoperability:

- Close air support
- Rendezvous
- Multiforce command and control
- Pinpoint cargo drop operations
- Search/rescue/evacuation operations

PPS

Most of the military applications make use of the very accurate PPS, although a few use SPS. The Department of Defense has established a policy in regards to civilian use of the PPS which specifies the requirements that must be met in order to grant civilian access to full GPS accuracies. The requirements indicate that access may be allowed if:

- It is in the national interest of the United States to allow access.
- The required accuracy cannot be achieved by other means (additional equipment or alternate methods).
- The security concerns of the GPS PPS are adequately provided for (who will use it and how it will be protected).

At this time it is assumable that, with few exceptions, the PPS is not available (or going to be available) for civilian use.

CIVILIAN APPLICATIONS

For civilian purposes, the Federal Radionavigation Plan states:

> The GPS SPS will provide a broad spectrum of civilian users with a sufficiently accurate PVT determination capability at a reasonable cost. Based on the joint DOD and DOT agreement documented in the FRP, an SPS position accuracy specification has been established for the civil community. Civil users will be able to determine their position to within 100 meters 2 drms (twice the distance rms) once the full GPS constellation is operational.

The applications for GPS in the civilian arena appears to be without limit. More and more uses are being found all the time (Fig. 4-6).

Marine use

Probably the largest current application of GPS by civilians, in number of users, is found in pleasure boating. Never has there been a single, one-stop source

Air navigation

Marine navigation

Land navigation

Recreation

Surveying

Search and rescue

Timing

Fig. 4-6. GPS civil applications.

for accurate navigation for marine purposes offering the accuracy and global coverage of GPS. Celestial navigation, although usable worldwide, lacks in exacting accuracy and can be crippled by weather conditions.

Various electronic, radio, and mechanical systems including radar, radio beacons and LORAN, and mechanical gyro systems have been used (and are still in current use). However, none can compare to the simplicity, coverage, and accuracy of GPS.

Only a few years ago the LORAN-C system became popular for marine navigation. It met a need with an interface that was simple to use, inexpensive, accurate, and reliable. However, today, more and more users are switching from LORAN to the GPS. Although a "current rage," GPS is recognized as a better system for many technical reasons, specifically including reduced atmospheric interference. Much of the current GPS marine hardware is based upon LORAN equipment development and design experience.

GPS for boaters travels the same equipment gamut that aviation equipment follows. From hand-held receiver/processors to moving-map displays, there is a device to meet nearly any need and a price for most pocket books (Fig. 4-7).

Boaters are using the SPS, which means typical 100 meter accuracy. To increase the accuracy, the USCG is installing differential GPS systems in many locations. The USCG DGPS is capable of increasing the accuracy to about 10 meters (39 feet), an increase of ten fold (Fig. 4-8). A diagram of the basic system is shown in chapter 2.

The increased accuracy afforded by DGPS allows boaters to navigate in harbors during periods of poor visibility while knowing their exact position, allowing them to avoid breakwaters, shoals, rocks, shallow areas, etc. and locate their desired destination (Fig. 4-9).

Other popular boater applications for GPS include the ability to return to a previous location. This may be for mooring purposes or to locate something specific:

- Objects of interest for diving
- Known fishing holes
- Rendezvous points

Fig. 4-7. An inexpensive, yet very accurate, marine-use GPS receiver that can be hand-held or mounted in place.

Fig. 4-8. By use of DGPS, accuracy down to 10 meters is achievable.

Magellan Systems

Fig. 4-9. With its 7-inch CRT, this Garmin GPS MAP 200 receiver/ processor can zoom in on details as small as 1 meter, and shows position by way of on-screen charts.

Garmin International

Accurate high-sea navigation is afforded all shipping and commercial water traffic via GPS, resulting in faster trips with more fuel efficiency. Currently, the equipment used on board ships rivals that installed on commercial airliners. GPS is integrated with the bridge system, providing direct coupling to auto-pilot and other navigation systems. Moving maps are commonplace and movement in harbors and on inland waterways has been made safer and faster.

When thinking of commercial shipping and GPS-based navigation, I must digress momentarily and wonder if the Exxon Valdez incident could have been prevented if GPS was in use?

A further feature of using GPS for maritime and boating navigation is a continuously known, accurate position in the event of a disaster, thereby reducing the time lost during a search.

Surface transportation

GPS will benefit surface modes of transportation in manners similar to other forms of transportation. Improved efficiency and safety are the two most prominent (Figs. 4-10 and 4-11).

One of the most ambitious applications projected for GPS is the Intelligent Vehicle Highway System (IVHS) technology. It encompasses automated highways and computer-aided vehicle guidance. Although not well known, IVHS is planned for limited use by the year 2000 with nearly a billion dollars already spent or allocated for its development and implementation.

Fig. 4-10. This FleetVision system provides location (left) and street mapping via StarView (right) information to provide efficient resource use.

Fig. 4-11. Application of a moving contour map for surface navigation.

The use of moving maps for automobiles is being developed by the Japanese Honda Motor Company and is available in Japan on a limited basis.

A small number of technically advanced trucking companies are using GPS as part of their scheduling system. Linked through a satellite communications system, each truck can automatically tell the home office where it is and what it is carrying. This information is then processed and trip efficiency examined. Further, knowing the locations of equipment on the road allows the trucking dispatcher to make more efficient use of resources when scheduling loads to be picked up and dropped off.

The rail system in the United States consists primarily of single sets of tracks with many trains sharing them. It is the function of the railroad traffic manager to schedule and clear trains to these tracks and provide for safe passage, therefore, the manager must have accurate position information of the various enroute trains to provide efficient use of the tracks, to prevent collisions, and reduce delays. Prior to GPS, everything depended upon schedules and timing, while exact train positions were often not known.

To facilitate the passing of opposing trains on the same track, sidings are used (a siding is a parallel track allowing one train to move from the active track while another passes). If the sided train arrived early, or the passing train was late, there was wasted time. If the passing train arrived early, and the train had not yet sided, there could be a collision. As the GPS system operates in real time, the traffic manager constantly knows the positions and status of all trains.

Public safety

GPS will assist in resource management in the realm of public safety. Dispatchers will be able to accurately determine where resources such as police cars, fire trucks, and rescue equipment are located at any given time. A computer-based history of this information can be used for later analysis.

When a police car suddenly fails to respond to a query from the dispatcher it is assumed the officer operating that car is in trouble and unable to acknowledge the call. In the past, the normal dispatcher response would be to call again and again. Eventually, other units would be dispatched to the last-known location of the unresponsive unit, hoping it would be located quickly. With GPS, the dispatcher knows the exact position of every car available or in use, via telemetry, and can send a unit to assist whenever necessary. GPS could well save an officer's life—just by providing a precise location to the dispatcher.

Fire departments will be able to use historic and map-based information giving locations of fire hydrants and other water supplies. Directions to a fire can be displayed on a computerized map on-board the fire apparatus, thus eliminating delays in arrival. This is particularly important in rural areas.

Using DGPS, individual fire fighters can locate themselves, or be remotely located, in the event they require help. No more searching an entire burning building when a fire fighter is missing. The location would be known and could be reported to those in command.

Equipment can be guided to the scene of a rescue and then to a medical facility by GPS, which will greatly reduce wasted time and inefficient travel.

Likewise, civil authorities responding to a natural disaster, such as an earthquake, will be able to easily locate underground pipelines and storage tanks as well as build a database of locations requiring further attention.

Geodesy

The mineral exploration and geophysical survey communities use GPS to accurately locate bodies of ore, potential petroleum-bearing areas, property lines, active earthquake faults, etc. very quickly.

Accurate positioning of oil-exploration equipment is essential for pinpointing promising oil-bearing geological formations from reflected seismic pulses. Oil-exploration companies are now using the SPS in conjunction with differential GPS techniques for this purpose. The alternative to GPS is an assortment of navigational systems, including the less convenient Transit system and shore-based transmitters—none of which offers the speed and accuracy of GPS.

Land survey and mapping can be done to accuracies of 5 cm (2 inches) and less. The equipment used receives L1 and L2 signals and uses signal phasing techniques for TOA computations and equation solving. Further, post-operation data processing is generally required for stored position/time information (Figs. 4-12 and 4-13).

Photogrammatry can accurately be done with only airborne equipment for precise locating of shot points.

Geodetic monitoring of formations is possible using GPS. Growth, movement, and deformation may all be monitored by the installation of fixed equipment. Accuracies are in the sub-meter range (less than an inch).

Explorations, forestry, and wildlife

GPS has proved to be an invaluable tool in the exploration of the Antarctic, an area where traditional navigation was always poor at best. A comment made about GPS in the Antarctic indicates that users can easily navigate directly to sites that are camouflaged from the air by large amounts of drifting snow. No longer are

Trimble Navigation

Fig. 4-12. A one-person survey team carrying all the equipment necessary for very precise positioning.

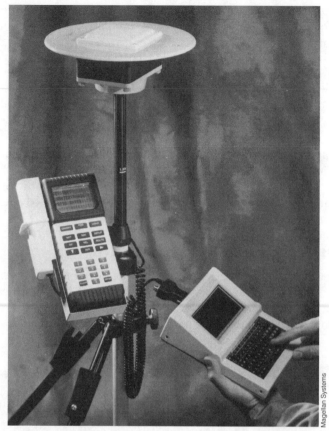

Fig. 4-13. Using the post-processing module (lower right), this Magellan NAV 5000 PRO, and its surrounding equipment, provides position information below the one-meter level.

near-location estimates used, followed by considerable eye-straining manual searching for signs of sites.

The applications of GPS in forestry include land survey, location of forests, estimates of forest content and size, and the mapping and building of logging roads. Assessments can be made of various animal (game) herd locations and sizes for hunting and control purposes.

Utility monitoring

Aerial monitoring of power lines and pipelines can easily be accomplished using GPS. Merely fly over the selected line and electronically mark (by pushing a single button on a receiver/processor) the location(s) of any problem areas noted.

After the observation flight is completed the information can be downloaded from the receiver for processing and future use.

Recreational use

Hikers are making use of GPS for its accuracy in outback areas. The precise locations given by a GPS receiver can be applied to maps, preventing hikers from becoming lost. Further, in the event of an emergency, hikers can give their exact locations to rescuers.

In the desert of the American Southwest, GPS provides positioning information that allows the return to the same place again and again, and for map use purposes.

Hunters will be able to go to very remote areas and have no fear of becoming lost. They will also be able to accurately return to any given point at a later time.

For the sports person, space age navigation is available to nearly everyone in nearly any adventure (Fig. 4-14).

In space

GPS has even found several major applications in space. Beginning with NASA's LANDSAT 4 in the early 1980s and continuing through today's TOPEX and Gravity Probe B programs, space-based GPS receivers demonstrate their capability for orbit determination. Use of precise time transfer from GPS has also played a significant role in the synchronization of ground-based spacecraft tracking networks.

Trimble Navigation

Fig. 4-14. This hang-glider pilot (with no power available what-so-ever) is using a hand-held GPS receiver for navigation.

INTERESTING EXPERIENCES

Many unique experiences in the use of GPS have resulted from the varied applications that are possible. Some track environmental problems, while others have ended arguments about mountain heights.

The National Park Service is using GPS in a study of the tortoise population of the Joshua Tree National Monument—the tortoise is an endangered species—and the University of Wyoming is tracking elk with a GPS-based telemetry system.

In June 1989 a GPS expedition scaled the tallest mountain in North America, Mt. McKinley, to determine its exact height. Using a self-contained Ashtech XII receiver/processor (Fig. 4-15), the exact height was determined as 20,306 feet, 14 feet below the 1954 trigonometric measurement. Other mountains measured using GPS include:

Mt. Fuji, tallest mountain in Japan, at 12,382 feet
Mt. Logan, Canada's tallest peak, at 19,546 feet
Mt. Everest, tallest in the world, at 29,022 feet

In 1992 GPS was used extensively by Esther Jacobson, Ph.D., during her exploration and archaeological journey to the Altay Mountains in the steppes of Siberia. Precise locations of many petroglyphs and ritual structures were recorded for future use.

GPS was used to accurately place the position of the Titanic, the famed ocean liner that sank after striking an iceberg.

GPS provides accurate system time at 16 primary nodes within the AT&T (the long distance company) network to prevent the degradation of digital customer services, including FAX, electronic data, video, and encrypted speech.

Tracking of high-altitude balloons during ozone depletion studies over the Arctic and Antarctic areas was done with GPS.

In 1993, during the resupply of Somalia, GPS provided the only means of accurate navigation for locating remote villages. Most of the villages look quite similar from the air and available maps were generally inaccurate by a matter of miles.

During the Great 1993 Flood of the upper Mississippi River Valley, GPS was used extensively to monitor areas covered with water. Flights were made on a daily basis over the flood areas and the locations of waters entered into the receivers. Later this information would be analyzed to indicate the floods progress, extent, and history.

Fig. 4-15. Ashtech XII GPS receiver used for precision static, kinematic, and pseudokinematic survey work.

5

Alternatives to GPS navigation

AIRCRAFT NAVIGATION is the process of piloting an aircraft from one place to another and includes position determination, establishment of course and distance to the desired destination, and computation of deviation from the desired track. Requirements for navigational performance are dictated by the phase of flight operations and their relationship to terrain, to other aircraft, and to the air traffic control process. There are two basic phases of air navigation: enroute/terminal and approach/landing.

Enroute/terminal

The enroute/terminal phase includes all portions of flight except that within the approach/landing phase. It contains subphases that are categorized by differing geographic areas, traffic densities, and operations.

Domestic enroute (high-altitude and low-altitude) routes are typically characterized by moderate to high traffic densities. Independent surveillance is generally available to assist in ground monitoring of aircraft positions.

Terminal is typically characterized by moderate to high traffic densities in transition from enroute to the approach/landing phase consisting of converging routes and changes in flight altitudes. Independent surveillance is generally available to assist in ground monitoring of aircraft positions.

Remote areas are special geographic or environmental areas characterized by low-traffic density and terrain where it has been difficult to cost-effectively implement comprehensive navigation coverage and which do not meet the requirements for installation of VOR/DME service or where it is impractical to install it. These include offshore areas, mountainous areas, and a large portion of the state of Alaska.

Operations between ground level and 5000 feet AGL (above ground level) are characterized by enroute flight operations below the 5000 feet AGL limit. Most helicopter operations as well as some fixed-wing operations are conducted in this subphase. This subphase typically has limited communication, navigation, and surveillance service because radio signals are easily blocked by terrain and buildings.

Area navigation (RNAV) is very satisfactory for low-traffic density operations, however, as traffic density increases, fixed low-altitude routes may be necessary to properly regulate traffic. Operations in metropolitan areas (high-density traffic) require the integration of both the enroute and terminal phases with nonprecision and precision approaches.

Approach/landing

The approach/landing phase is that portion of flight conducted immediately prior to touchdown and is generally done within 10 nautical miles of the runway. Two subphases may be classified as nonprecision and precision approaches to landing.

Nonprecision approach

Minimum safe altitude, obstacle clearance area, visibility minimums, final approach segment area, etc., are all functions of the navigational accuracy available and other factors. The unique features of RNAV for nonprecision approaches are specified in FAA Advisory Circulars No. 90-45A, "Approval of Area Navigation Systems for Use in the U.S. National Airspace System;" No. 20-130, "Airworthiness Approval of Multi-Sensor Navigation Systems in U.S. National Airspace System (NAS) and Alaska;" and 20-121A, "Airworthiness Approval of the LORAN-C Navigation System for Use in U.S. National Airspace (NAS) and Alaska." All are available from the FAA.

Precision approach and landing

Precision approach and landing navaids provide vertical and horizontal guidance and position information. The instrument landing system (ILS) and the proposed microwave landing system (MLS) are both precision systems.

Present navaid service

The GPS is relatively new to the general-aviation user as a navigational aid. More familiar are the VORs, NDBs, DME, LORAN-C, and the still incomplete MLS. Although each has been in use for sometime, and as of today, are in everyday use, they will ultimately be replaced by GPS.

VOR—the FAA's VHF (very high frequency) based current point-to-point navigation system.

DME—distance measuring equipment, generally co-located with a VOR that provides position information in terms of miles.

RNAV—area navigation is a computerized VOR/DME application not requiring use of designated airways.

NDB—nondirectional radiobeacons used for general air navigation (in less developed regions of the world) and as a part of many instrument approaches.

LORAN-C—a computer-based positioning system, primarily designed for marine use (and maintained by the USCG), now in common use in the general-aviation fleet.

ILS—VHF precision approach and landing system.

MLS—microwave landing system, a proposed replacement for the current ILS system.

Recall in chapter 1, where a small portion of the Federal Radionavigation Plan was quoted:

> . . . many existing navigation systems are under consideration for replacement by GPS beginning in the mid-to-late 1990s. GPS may ultimately supplant less-accurate systems such as LORAN-C, Omega, VOR, DME, TACAN, and Transit, thereby substantially reducing federal maintenance and operating costs associated with these current radionavigation systems.

The proper use of the tax dollar and citizen demand for less waste will require this portion of the FRP to be followed. Further, various short-comings of the older systems make them obsolete, or less useful, less accurate, and less safe than GPS.

THE VOR-, DME-, AND TACAN-BASED SYSTEMS

An examination of the current navigation systems supporting general aviation shows how each works and to what extent GPS can replace them. The three systems that provide the basic guidance for enroute and terminal air navigation in the United States are VOR, DME, and TACAN. System-provided information consists of azimuth and distance relative to the facility.

The VOR system

The VOR system serves as the primary means of navigation for civil aviation in the national airspace system. In use since the 1950s, it is a FAA-owned and operated navigation system and was developed as a replacement for the low-frequency radio range (A–N ranges) (Fig. 5-1). The system's purpose is to provide a bearing from a VOR transmitter to an aircraft.

The FAA operates more than 900 VOR, VOR/DME, and VORTAC stations including 150 VOR-only stations. A collocated DME (distance measuring equipment) provides the distance from the aircraft to the DME transmitter. At most sites, the DME function is provided by the TACAN system which gives azimuth and distance guidance to military users. Such combined facilities are referred to as VORTACS.

The VHF radio band limits propagation to line-of-sight, generally giving a usable reception distance of 25 to 200 nm. Actual usable range distances will vary according to the terrain, location of the transmitter, altitude of the airplane and class of the navigational aid. VOR operating frequencies are between 108.0 and 117.95 MHz. The three classes of VORs are:

T (terminal) VOR—normal usable range is 25 nm at altitudes of 12,000 feet or less. Transmitter output power is about 50 watts.

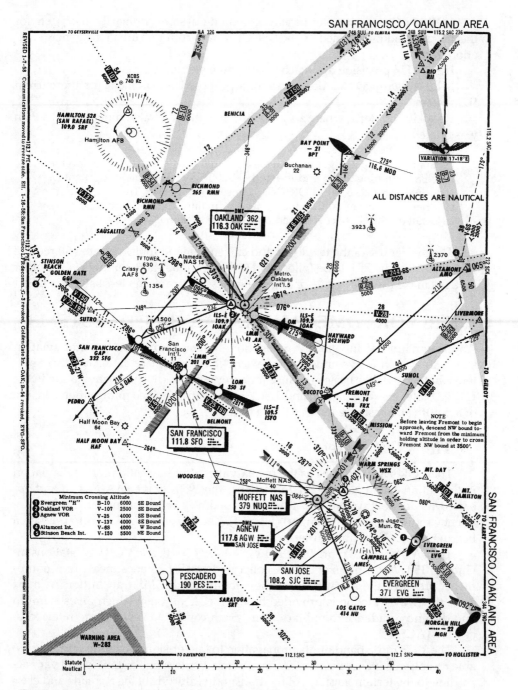

Fig. 5-1. In the center of this chart notice the unusual indicators for 124, 200, 304, and 020 degree radials at Oakland. They indicate a low-frequency A & N radio range (circa 1958), where the pilot kept to the center by listening to a combination of Morse code As and Ns. Illustration only and not for navigation. (Copyright 1958 Jeppesen Sanderson, Inc., reproduced with permission of Jeppesen Sanderson, Inc.)

L (low altitude) VOR—providing a usable range of 40 nm at altitudes of 18,000 feet or below.

H (high altitude) VOR—giving coverage of 100 nm at altitudes above 5000 feet. Above 20,000 feet the range will approach 200 nm. These stations have a transmit power of approximately 200 watts.

How the VOR works. The VOR transmits two signals. The first is called the reference signal and has a constant phase at all points around the transmitter. The other is segmented into the 360 degrees (radials) of the compass by electrically rotating it around the first. This movement causes a change of phase at all points other than magnetic north. The two components are in phase at magnetic north (VOR stations are oriented to magnetic north) and out of phase, to a varying degree, in all other directions.

The airborne part of the VOR system is a receiver that compares the phase difference of the VOR's two-part signal and determines on which radial—compass direction or azimuth from the VOR—it, the receiver in the airplane, is located. The radial information is typically displayed on a course deviation indicator, called a CDI (Figs. 5-2 and 5-3) or, in the case of digital numeric displays, numerically (Fig. 5-4).

VOR radial information. Knowing the radial of the VOR indicates only the direction the aircraft is from the VOR station. It does not indicate distance. There-

Fig. 5-2. IND350A CDI for VOR use.

Fig. 5-3. Electronic CDI showing digital information and the familiar "cross needles."

Fig. 5-4. The active NAV channel indicated on this MK-12D is 115.20 MHz and the plane is on the 168 degree leg of the VOR.

fore, you could be anyplace along the radial, a distance up to 200 nm. To place the radial information on a map, such as an appropriate aeronautical chart, draw a line extending from the VOR along the radial indicated by the VOR receiver. The line is called a LOP—line of position.

Generally, VORs provide accurate information when the ground transmitter and airborne receiver are operating within specifications. There are, however, areas where VOR signals can become unreliable due to geographical features, such as mountains, and in the immediate vicinity of the station itself where there is a cone of silence.

VOR positioning. If two or more VOR stations are used, positioning can be made by locating the point at which the received radials cross. This is easily visualized by drawing LOPs from each of the VOR stations (Fig. 5-5). The point of crossing (intersection) is the aircraft's position (also called a fix).

Accuracy of positioning requires simultaneous readings from two VORs, as a VOR receiver can tune only a single station at a time. The most accurate fixes result from radials that are 90 degrees opposed.

Integrity. VOR provides system integrity by removing a signal from use within ten seconds of an out-of-tolerance condition detected by an independent monitor.

VOR strengths and weaknesses. The VOR system has been around for nearly 40 years. It serves general, commercial, and military aviation in an efficient manner and is a very familiar technology. The system offers specific strengths and weaknesses:

Strengths of VOR: User equipment is relatively inexpensive, generally under $2000 for a competent single receiver and CDI installation. The VOR system is passive—allowing many simultaneous users, nearly 100 percent CONUS (continental U.S.) coverage; ease of use via CDI or digital display information (radial display), provides easy wind-drift correction without calculations; individual aircraft installations may be checked on a VOT (VOR test facility) or by checking over a known radial checkpoint that is listed in the Airport/Facilities Directory; and VOR stations can be used for voice communication with an associated Flight Service Station (FSS) . Some facilities provide TWEBs (transcribed weather broadcast) on VORs.

Weaknesses of VOR: VOR range is typically limited to 130 nm at best, and often less, for most general aviation uses, system accuracy is approximately one degree. This equates to about three miles at 130 nm range. Further, accuracy checks of VOR receivers require only ±4 degrees (ground check) or ±6 degrees (airborne check). With this in mind, a total error of nearly 20 miles is possible at maximum

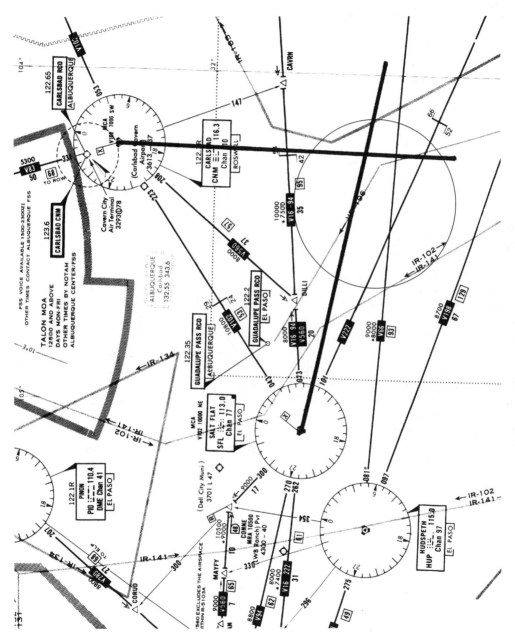

Fig. 5-5. The LOP from SFL intersects the LOP from CNM giving the airplane's position or fix.

range, station identification on some stations is by means of Morse code letters sent as part of the station's transmission, TO and FROM information displayed on the CDI can be confusing, and no indication of ground speed may be directly determined.

Distance measuring equipment

Distance measuring equipment—DME—works with the VOR to provide distance and calculated groundspeed information. Typically, DME stations are collocated with VOR stations, however, this is not a requirement. Stand alone DME stations do exist, but most VORs have DME equipment collocated with them, usually in the form of a VORTAC.

How DME operates. DME is an active system, requiring receivers and transmitters at both the station and user ends. The DME system is initiated by a signal from the airborne unit. The frequencies of DME operation are in the UHF (ultra-high frequency) range and are paired with VOR frequencies, providing ease of use.

The airborne DME transmits paired pulses and the VORTAC station responds by sending its own paired pulses back to the aircraft's DME equipment. The airborne unit observes the time interval between initial transmission and final reception of the pulses and determines the distance (range) via calculation (Fig. 5-6).

Some distance measuring equipment is capable of determining ground speed and ETA (estimated time of arrival)/ETE (estimated time enroute) based upon distance observations and calculations. Note that for an accurate ground speed calculation, the plane must be going directly to or from the VORTAC. Further, as the plane gets close to the station, slant-range error can cause a gross distortion of the groundspeed. Slant-range error is caused by the distance down to the VORTAC station being included in the range distance.

It is possible to make position fixes based upon distance ranging in a manner similar to GPS TOA ranging. The procedure is called rho-rho navigation (rho is the Greek letter used as the navigational symbol for distance), however, has little application in general aviation (Fig. 5-7).

Accuracy and range of DME. DME accuracy depends upon the specific unit design, however, is usually within three percent (4 nm at 130 nm range). Note that many units actually deliver .2 nm accuracy on a routine basis.

Because of the operating frequencies (960 to 1213 MHz), DME has a line-of-sight limitation (similar to that of VHF), which restricts coverage to 30 nm or less in terminal areas at terminal altitudes. At altitudes above 5000 feet the range will increase to more than 100 nm. Enroute stations radiate at 1000 watts of power and terminal DMEs at 100 watts.

Integrity. DME provides system integrity by removing a signal from use within ten seconds of an out-of-tolerance condition detected by an independent monitor.

Strengths and weaknesses of DME. The instrument-panel mounted DME may be the simplest piece of navigation equipment there is to operate. Most have automatic tuning which is slaved to the VOR receiver to make use of the paired frequency scheme.

DME is very accurate, however it is limited in application except when coupled with VOR azimuth information.

Fig. 5-6. The distance indicated is 97.9 nm, the calculated speed is 120 kts, and the ETE is 48 minutes on this DME 890.

Narco Avionics

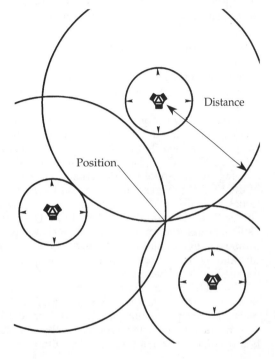

Fig. 5-7. Rho-Rho navigation is based upon intersecting circles of DME measured distance. Fix is at the common point of all distances measured.

Distance

Position

DME Strengths: Ease of operation, wide availability of VOR/DME and RTAC stations, and the relative accuracy is excellent.

DME Weaknesses: Slant-range errors (distance and ground speed) and the capacity of DME is limited to 110 simultaneous users, although actual overload is rare.

TACAN

TACAN is a UHF radionavigation system that provides a pilot with relative bearing and distance to a beacon on the ground, ship, or to specially equipped aircraft. It is a short-range navigation system used primarily by military aircraft.

TACAN stations are often collocated with the civil VOR stations (VORTAC facilities) to permit military aircraft to operate in civil airspace.

TACAN users. There are approximately 14,500 aircraft presently equipped to determine bearing and distance to TACAN beacons, consisting primarily of navy, air force, and to a lesser extent, army aircraft. Additionally, allied and third-world military aircraft use TACAN extensively. NATO has standardized on TACAN until 1995.

Limitation to use. Because of propagation characteristics of the UHF frequencies used (960 to 1215 MHz) and radiated-power limits, TACAN is limited to line-of-sight or to approximately 200 miles at higher altitudes (above 18,000 feet). As with VOR/DME, special consideration must be given to locations of ground-based TACAN facilities, especially in areas where mountainous terrain is involved, due to its line-of-sight coverage.

Integrity. TACAN provides system integrity by removing a signal from use within ten seconds of an out-of-tolerance condition detected by an independent monitor.

FRP statement regarding DME, VOR, and TACAN:

The overall VOR/DME system is protected by international agreement until 1995. It is expected to remain in service through the year 2010. If an alternate system such as LORAN-C or GPS should prove acceptable to the international aviation community as a replacement for VOR/DME, a significant level of implementation would not occur until the late 1990s. It would require a substantial period beyond that before VOR/DME phase-out could be accomplished.

The DOD VOR/DME operational concept is to maintain present system coverage until a suitable replacement is available. Present plans for expansion of the VOR/DME system are limited to site modernization or facility relocation. GPS is the planned replacement for DOD VOR/DME and VORTAC facilities. This transition started in 1988. The target date for phase-out of the DOD requirement for VOR and VOR/DME is the year 2000. In the case of a military VORTAC site that has developed an appreciable civilian-use community and is due for phase-out, transfer of operational responsibility to the DOT will be discussed between DOD and DOT.

The DOD requirement for and use of land-based TACAN will terminate when aircraft are properly integrated with GPS and when GPS is certified to meet RNP (required navigation performance) in national and international controlled airspace. The target date is the year 2000. The requirement for shipboard TACAN will continue until a suitable replacement is operational.

RNAV

RNAV, meaning *area navigation,* is generally applied to the VORTAC-based system. The heart of an RNAV unit is its computer, which works in conjunction with an airborne VOR receiver and DME, sometimes slaved and others as a single integrated unit.

The function of the RNAV equipment is to electronically relocate a VORTAC from its real location to a user-selected location. The user-selected location is called a waypoint. The waypoint is defined as a radial and distance combination. For example: Plymouth, Vermont, is located on the 265-degree radial of the Lebanon VOR/DME at a distance of 24 miles.

The VOR/DME's frequency and the radial/distance information can be programmed into the RNAV unit to create a waypoint (or phantom VOR/DME). In this case, the waypoint could be called Plymouth. The RNAV unit can then navigate to and from this user waypoint as well as if it were a real VOR/DME, including providing radial and DME reports.

RNAV uses the rule of trigonometry for its calculations that states, "If two sides of a triangle and the angle they form are known, then the third side can be determined."

Strengths and weaknesses of RNAV. The RNAV units allow relatively accurate navigation to precise locations that do not have navaids (VORTAC or VOR/DME) by means of user programmed waypoints. Multiple waypoints can be used.

Strengths of RNAV: Allows random routings and waypoints, not just along standard airways or between VORTACS and is excellent for locating specific geographical locations.

Weaknesses of RNAV: Waypoints must be within the VORTAC's range and requires the use of VORTAC or VOR/DME (cannot be used with VOR alone).

At this time it is safe to say that RNAV will be usable until the demise of the VOR (VORTAC and VOR/DME) system.

ILS

ILS—instrument landing system—provides precision vertical and horizontal navigation (guidance) information to aircraft during approach and landing. Associated marker beacons or DME equipment identify the final approach fix, the point where the final descent to the runway is initiated.

ILS is the standard civilian aviation landing system in the U.S. and the international standard for aircraft operating under IFR conditions. Since its introduction in the 1940s, it has been installed in steadily growing numbers throughout the world. Part of its attractiveness to aircraft owners lies in the economy of avionics costs. Since the ILS localizers and VOR stations operate in the same frequency band, common VHF receivers are used. Glideslope transmissions are UHF.

Military services use ILS at fixed bases in the U.S. and overseas. Special systems are used to meet unique military requirements, including shipboard operations.

ILS is everywhere

In 1992, there were 974 ILS sites in the United States. Eventually, nearly 1100 ILS sites will exist. In addition, there are approximately 165 ILS facilities operated by DOD in the United States.

Federal regulations require U.S. air-carrier aircraft to be equipped with ILS avionics and ILS is commonly used by general-aviation aircraft. Since ILS is the International Civil Aviation Organization standard landing system, it is extensively used by air-carrier and general-aviation aircraft of other countries.

Based on a 1990 user survey, the number of civilian aircraft equipped with ILS is estimated to be 125,000 (Fig. 5-8).

How ILS works. The localizer facility and antenna are typically located 1000 feet beyond the stop end of the runway and provide a VHF (108 to 112 MHz) signal. The glideslope facility is located approximately 1000 feet from the approach end of the runway and provides a UHF (328.6 to 335.4 MHz) signal.

Fig. 5-8. IND351A CDI with glide slope indicator.

Marker beacons are located along an extension of the runway centerline and identify particular locations on the approach (Fig. 5-9). These 75 MHz beacons are included as part of the instrument landing system: an outer marker at the initial approach fix—typically four to seven miles from the approach end of the runway—and a middle marker located approximately 3500 feet from the runway threshold. The middle marker is located so as to note impending visual acquisition of the runway in conditions of minimum visibility for Category I ILS approaches. An inner marker, located approximately 1000 feet from the threshold, is normally associated with Category II and III ILS approaches.

Accuracy and integrity. Course alignment (localizer) at threshold is maintained within 25 feet. Course bends during the final segment of the approach do not exceed .06 degrees (2 sigma). Glide slope course alignment is maintained within 7.0 feet at 100 feet (2 sigma) elevation and glide path bends during the final segment of the approach do not exceed .07 degrees (2 sigma).

ILS reliability approaches 100 percent. However, terrain and other factors may impose limitations upon the use of the ILS signal. Special account must be taken for terrain factors and dynamic factors such as taxiing aircraft, which can cause multi-path signal transmissions.

In some cases, to resolve ILS-placement problems, use has been made of localizers with wide aperture antennas and two-frequency systems. In the case of the glideslope, use has been made of wide aperture, two-frequency image arrays and single-frequency broadside arrays to provide service at difficult sites.

ILS provides system integrity by removing a signal from use when an out-of-tolerance condition is detected by an integral monitor.

Limitations of ILS. Limitations of the ILS manifest themselves in three major areas:

1. Performance of individual systems can be affected by terrain and manmade obstacles such as buildings and surface objects like taxiing aircraft and snow banks.
2. The straight-line approach path inherent in ILS constrains airport operations to a single-approach ground track for each runway.

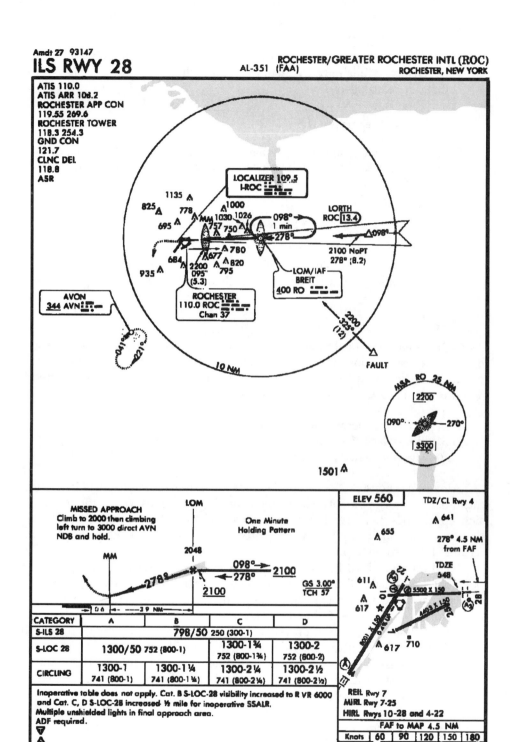

Fig. 5-9. Note the LF marker beacons along the approach and their positions relative to the projected flight path. (Illustration only and not for navigation.)

3. Even though the new 50 kHz frequency spacing will eventually double the ILS channel availability, frequency saturation limits the number of systems that can be installed.

FRP statement regarding ILS:

ILS is the standard civil landing system in the U.S. and abroad, and is protected by ICAO (International Civil Aviation Organization) agreement to January 1, 1998.

OTHER NAVAIDS

Other navaids are available for use by the pilot as alternatives to the VOR-based system. Some are very simple in installation (from the FAA point of view) while others are navigation aids/systems not initially designed for aviation purposes.

NDBs

NDBs—nondirectional radiobeacons—are not used for general air navigation in the United States, however, they are an important part of many instrument approaches. Presently, there are 1575 low- and medium-frequency aeronautical nondirectional beacons distributed as follows:

FAA-operated Federal facilities: 728
U.S. Military facilities: more than 200
Non-federally owned facilities: 847

Aircraft use nondirectional radiobeacons as compass locators to aid in finding the initial approach point of an instrument landing system as well as for nonprecision approaches at low-traffic density airports without conventional VOR approaches. Further, many of the nondirectional beacons are used to provide weather information to pilots.

NDB theory. All AM radio stations that transmit nondirectional signals are potential nondirectional beacons, including AM commercial-broadcast stations. NDBs for air navigation operate in the low- and medium-frequency (L/MF) radio bands from 190 to 415 and 510 to 535 kHz.

NDB transmitters vary in power from less than 25 watts output up to 2000 watts with useful ranges varying from 15 nm to several hundred nm. Actual range depends on transmitter output power and atmospheric conditions. The nominal range of the most powerful NDBs is 75 nautical miles, however, they can often be heard several hundred miles away.

The equipment used in an airplane to navigate via NDBs is the ADF (automatic direction finder) (Fig. 5-10). It consists of an AM receiver, two antennas (sense and loop), and an indicator.

Determination of the signal's direction is made by electronically rotating the antenna and noting the relative received signal strength. The resulting output is

Fig. 5-10. KR 87 TSO ADF receiver.

AlliedSignal Inc., General Aviation Avionics

the direction, or bearing, to the transmitting station (NDB), generally displayed on a compass card by means of a pointer (Fig. 5-11). NDB identification is made by Morse code identifiers.

Types of NDBs. There are four categories of NDBs, depending upon location and use:

- Compass locator, which is either at the outer or middle marker on an ILS approach. They are low-powered transmitters designed as approach orientation aids.
- NDB approach facility, found on or near airports where NDB is the primary approach aid. The intended operational range is 25 nm.
- Enroute airway beacons are common to parts of the world other than the United States, except Alaska where they provide some enroute navigation services. They generally have an operational range of 50 miles and are used for navigating airways.
- The high power beacon provides navigational service for over-water routes and between shore and island-based facilities. Generally, their usefulness has been replaced by more sophisticated navigation systems.

ADF operation. ADF operation is very simple because the airborne unit does nearly all the work for the pilot. Modern ADF receivers are digitally tuned and require only the appropriate NDB's frequency to be entered.

Early NDBs were called homing beacons. Homing to a beacon simply means pointing the aircraft in the direction of the station and flying to it. Basically, ADF compass-indicator needles always point in the direction of the station (beacon).

AlliedSignal Inc., General Aviation Avionics

Fig. 5-11. ADF directional indicator.

Merely turn the aircraft until the indicator's needle points to the top of the indicator, thus pointing the aircraft directly toward the beacon and homing to it.

The homing method is simple, however inefficient, as any crosswind will push the airplane to one side or the other of the direct course to the NDB, causing the indicator needle to move to one side or the other of the direct course. The pilot will have to make corrections again and again to the course flown, with a resulting curved path actually being flown.

The repeated course changes made to correct for the crosswind merely amount to wind compensation. In essence, the airplane becomes a weather vane swinging into the wind about the NDB. The only time this situation will not exist is in a straight on headwind/tailwind or a no-wind condition.

To overcome the deficiencies of homing, ADF operation uses tracking. Tracking requires a constant wind correction angle to cancel the drift caused by the crosswind. Various methods of tracking have been developed, each of which is related to the type of compass-indicator card in use. There are three types of compass indicator cards:

Fixed card—displayed bearing is relative to the longitudinal axis of the aircraft (heading of the aircraft).

Rotatable card—displayed bearing is magnetic, as the compass card can be rotated to agree with aircraft's heading.

Radio magnetic indicator (RMI)—magnetic bearing with continuous gyro-stabilized heading information supplied to the compass card.

Limitations of using NDBs. Most limitations of the NDB system relate directly to the propagational factors of the radio frequencies used. Specifically, the L/MF radio band is plagued with many propagational drawbacks (Figs. 5-12 and 5-13).

Ground wave is stable, reliable, and may be heard for several hundred miles. The actual distance depends on station-transmitter output power and frequency. The lower the frequency and the higher the power, the stronger the ground wave.

Direct wave is line-of-sight, just like a VOR signal, and can be received up to a maximum of approximately 200 nautical miles at the highest altitudes normally used by general aviation.

Sky wave is a direct wave that has been reflected back to earth by the ionosphere. Reflections can be multiple when propagation conditions favor this effect (generally during the hours of darkness). The point (or points) at which the sky wave returns to earth may be hundreds or thousands of miles away from the source. The higher the frequency, the greater the distance between reception points.

Skip zone is a gap in coverage between the ground wave and the sky wave. For example: the ground wave coverage of an NDB extends out 180 nm. The sky wave returns at 900 nm. The area between the end of the ground wave and the sky wave return is called the skip zone (the signal skips over it). The exact length of the skip zone depends on the frequency of the signal, the transmitter output power, and on various atmospheric and ionospheric factors. The zone is not easily predictable and can severely limit the reliability of NDB use.

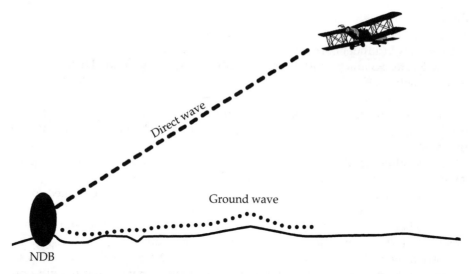

Fig. 5-12. Low frequency propagation of the ground wave and direct wave.

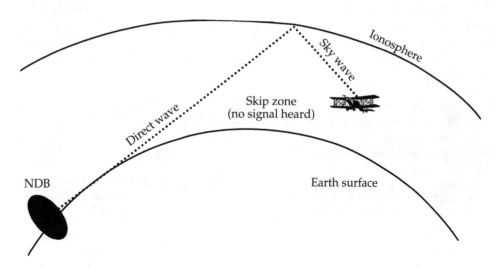

Fig. 5-13. Low frequency propagation of the sky wave showing the skip zone.

During the transition period from daylight to darkness, conditions of the ionosphere are changing—causing changes to NDB signal propagation. The result can be overlap of the ground wave by the sky wave, causing erratic indications on the part of the ADF.

Fading is a phase condition when simultaneous reception of ground wave and sky wave elements of the same signal coincide or conflict with one another. The results can effect ADF accuracy.

Thunderstorms, producing static interference from lightning, can cause erroneous or erratic needle indications. In most cases the ADF is unusable near heavy thunderstorm activity (often meaning within hundreds of miles). There is no real solution to the problem, except to use means other than NDBs for navigation during periods of thunderstorm activity.

LOPs and accuracy. LOPs (lines of position) can be determined using NDBs as easily as with VOR radials, however, the accuracy will be less with NDBs.

Bearing accuracy is on the order of 3 to 10 degrees and the lesser (3 degree) accuracy requires that the ADF be calibrated before it is used for navigation by comparing radio bearings to accurate bearings obtained visually on the transmitting antenna.

For aviation use the required accuracy is stated in terms of permissible needle swing: 5 degrees on approaches and 10 degrees in the enroute area.

Integrity. Out-of-tolerance conditions for aviation beacons are limited to output power reduction below operating minimums and loss of the CW identifier. The radiobeacons used for nonprecision approaches are monitored and will shut down within 15 seconds of an out-of-tolerance condition.

Strengths and weaknesses. The successful use of NDBs is subject to many signal-propagation factors beyond the control of the pilot, making it difficult to predict any standard of accuracy.

Strengths of NDBs: Ground-based equipment is inexpensive to install, operate, and maintain, therefore is used in places where no other navaids are available.

Weaknesses of NDBs: The nondirectional-beacon system is not as accurate as the VOR system and the radio signals are susceptible to propagational and atmospheric degradations.

As an instrument-approach aid, the NDB is far less accurate than the VOR. NDBs should not be relied upon unless no other navaid is available and then only used with caution.

FRP statement regarding NDBs:

During the next 10 years, planned FAA expenditures for beacons will be limited to the replacement of components, modernization of selected facilities, and an occasional establishment or relocation of an NDB used for ILS transition.

Aeronautical radiobeacons provide low-cost navigation and will remain part of the radionavigation systems mix through the year 2000. At this time, the probability of change beyond the year 2000 cannot be accurately predicted.

LORAN-C

LORAN-C is a computer-based navigational system installed and maintained outside the FAA navaid system. Originally intended for marine use only, LORAN-C has found solid application within aviation, both military and civilian. It is operated and maintained by the United States Coast Guard. LORAN-C is approved as a supplemental air navigation system and is approved for nonprecision approaches at certain airports.

Using a receiver/processor, similar in design to the more sophisticated GPS receiver/processor, LORAN provides accurate positioning and, via processing,

Fig. 5-14. KLN 88 panel-
mounted LORAN-C receiver
showing distance and route to
KATL information.

AlliedSignal Inc., General Aviation Avionics

Fig. 5-15. Voyager hand-held
LORAN-C receiver displaying
current location in geographical
coordinates.

bearing/distance, ground speed, ETE, and ETA information. The waypoint means
of location is prominent in LORAN usage (Figs. 5-14 and 5-15).

How LORAN-C works. LORAN is based upon chains of transmitting stations. A chain consists of a master station and two or more secondary stations.
They transmit signals in the LF (low frequency) radio band at about 100 kHz. At
these frequencies, propagation generally follows the contour of the earth's surface
and therefore can be received at great distances—hundreds or thousands of miles.
This characteristic allows use well beyond the line-of-sight limits of VHF.

The LORAN receiver selects the optimum chain for your area, then listens to
the pulse signals emitting from the master and secondary stations. Using an internal timing device, the receiver/processor measures the difference in signal arrival
time for the master and secondary signals. It then determines a pair of computerized LOPs, giving the fix in latitude and longitude coordinates as the aircraft's present position.

LORAN signal propagation is roughly six microseconds per nautical mile over
sea water. This speed becomes less and variable over land. The effect is called ASF
(additional secondary phase factor) and is compensated for by mathematical (or
database) means internally in the LORAN unit.

As with GPS, LORAN requires good signal geometry to obtain accurate fixes
and provide precise calculated information. Generally, the LORAN receiver will
look after these chores automatically.

Using LORAN. As the airplane travels along, the LORAN receiver/processor can determine position and with further positioning, determine speed and direction of flight. Generally, LORAN units can provide the user with bearing information based on the current waypoint, range to the waypoint, track (over ground), drift angle, and an assortment of time calculations.

LORAN receivers use digital-display systems either LCD (liquid crystal display) or LED (light emitting diode) matrixes for showing operator input and calculated data output.

Usable accuracy is measured in two ways for LORAN: repeatable and absolute. Repeatable accuracy refers to returning to a specific previous position. Absolute accuracy means the ability to determine a position independently.

Repeatable accuracy: 18 to 90 meters
Absolute accuracy: .1 to 2.5 nm

Individual case accuracy is a product of distance from the chain and geographical terrain, along with receiver design and the SNR (signal to noise ratio). Typically, this equates to 200 to 400 feet under operational conditions.

Many current LORAN receivers make use of extensive databases providing locations of airports and navaids.

LORAN instrument approaches are nonprecision (no glide slope) and are few and far between. Additionally, any use of LORAN for IFR is restricted unless the unit is IFR certified. Most are VFR only and cannot be used legally as primary navigation equipment under IFR conditions.

LORAN-C signal monitors have been installed throughout the national airspace system to support the use of LORAN-C as a nonprecision approach aid. The monitors are operated and maintained by the FAA. Each monitor will provide both long-term signal data for use in the prediction of signal corrections at individual airports and the status of LORAN-C signals for the local area.

Integrity. LORAN-C stations are constantly monitored to detect signal abnormalities that would render the system unusable for navigation purposes. The secondary stations "blink" to notify the user that a master-secondary pair is unusable. Blink begins immediately upon detection of an abnormality. The USCG and the FAA are also developing automatic blink equipment and concept of operations based on factors consistent with aviation use.

Strengths and weaknesses. LORAN-C was a great step when accepted for use in aviation. It allows accurate positioning in areas not served by the VHF VOR system. A well-designed receiver can eliminate the sky wave/ground wave interference problem and compensate when using the far-traveling sky wave.

Most units, due to their data processors, output more information than the pilot can use at any single point in time.

Strengths of LORAN-C: Much greater usable range than VHF VOR signals, data-processor based system provides use of waypoints, airborne equipment is relatively inexpensive, and no additional ground based equipment is required.

Weaknesses of LORAN-C: The SNR (signal to noise ratio) can be adversely affected by low signal strength or precipitation static. Sometimes LORAN is not usable in rain.

FRP statement regarding LORAN:

The LORAN-C system now serves the CONUS, its coastal areas, and certain parts of Alaska. Based upon demands from the aviation community, the FAA and USCG jointly sponsored expansion of the LORAN-C system to complete coverage over the United States (May 1991), providing reliable and accurate enroute and nonprecision approach navigation capabilities to a greater number of airports.

Although LORAN is expected to remain part of the radionavigation mix through the year 2015, the DOD requirement for the LORAN-C system will end December 31, 1994.

Note: DOD termination of use will not affect civil-aviation use of LORAN-C in the CONUS. The FAA has designated LORAN-C as a supplemental system in the National Airspace System and will fully implement it in the NAS by approving nonprecision approaches at selected airports that have adequate LORAN coverage. Toward that end, FAA has deployed 196 local LORAN-C monitors throughout the NAS to provide calibration values required for nonprecision approaches. The FAA and the USCG are also developing automatic blink equipment and a concept of operations to support nonprecision approaches in the NAS.

MLS

ICAO has selected the microwave landing system—as the international standard precision approach system, with implementation targeted for 1998. It is planned to replace ILS in national and international civil aviation. At this time, from a possible installation point of view, the FAA and DOD plan to have MLS collocated with ILS to minimize the transition impact.

The MLS is a joint development of DOT, DOD, and NASA under FAA management. Its purpose is to provide a civil/military, Federal/nonfederal standardized approach and landing system with improved performance compared with existing landing systems.

MLS does not have the geographical feature placement problems of ILS and offers higher accuracy and greater flexibility, permitting precision approaches at more airports.

How MLS works. MLS transmits signals that enable airborne units to determine the precise azimuth angle, elevation angle, and range. The technique chosen for the angle function of the MLS is based upon time-referenced scanning beams (TRSB). All angle functions of MLS operate in the 5.00 to 5.25 GHz band. Ranging is provided by DME operating in the 960 to 1215 MHz band. An option is included in the signal format to permit a special purpose system to operate in the 15.4 to 15.7 GHz band.

Accuracy and integrity. The azimuth accuracy of MLS is 13.0 feet at the runway threshold approach reference datum and the elevation accuracy is 2.0 feet. The lower surface of the MLS beam crosses the threshold at 8 feet above the runway centerline. The flare guidance accuracy is 1.2 feet throughout the touchdown zone and the DME accuracy is 100 feet for the precision mode and 1,600 feet for the nonprecision mode.

The MLS signals are generally less sensitive than ILS signals to the effects of snow, vegetation, terrain, structures, and taxiing aircraft. This allows the reliability of this system to approach 100 percent.

MLS integrity is provided by an integral monitor. The monitor shuts down the MLS within one second of an out-of-tolerance condition.

Disposition of MLS. As of this writing (1994), due to fiscal requirements and as a practical matter, the FAA is being compelled to re-examine its demand for the MLS and for the hungry budgets necessary for its development.

Although intended as a very accurate instrument-approach system, it is now felt that MLS equates only to great expense for hardware (ground and air based) and systems maintenance. Further, many aviation professionals and experts indicate that the system is doomed to obsolescence before it is ever placed into general use. Their recommendation is for increased implementation of GPS based instrument landing methods.

To further complicate matters, the U.S. has entered into various international agreements that may require the completion and implementation of MLS, even if obsolete.

CIVIL AIR RADIONAVIGATION REQUIREMENTS

Aircraft navigation can follow VFR (visual flight rules) and require a minimal amount of electronic navigation equipment or be under IFR (instrument flight rules) and require considerable electronic navigation equipment.

Navigation is only one function of the pilot, therefore it must not be allowed to overload the pilot and interfere with other flying duties. The following are basic requirements for the current and future aviation navigation system. The words "navigation system" mean all of the elements required to provide the necessary navigation services for each phase of flight. As stated in the FRP:

a. The navigation system must be suitable for use in all aircraft types which may require the service without limiting the performance characteristics or utility of those aircraft types; e.g., maneuverability and fuel economy.

b. The navigation system must be safe, reliable, and available; and appropriate elements must be capable of providing service over all the used airspace of the world, regardless of time, weather, terrain, and propagation anomalies.

c. The integrity of the navigation system, including the presentation of information in the cockpit, shall be as near 100 percent as is achievable and, to the extent feasible, should provide flight decwarnings in the event of failure, malfunction, or interruption.

d. The navigation system must have a capability of recovering from a temporary loss of signal in such a manner that the correct current position will be indicated without the need for complete resetting.

e. The navigation system must automatically present to the pilot adequate warning in case of malfunctioning of either the airborne or source element of the system. It must assure ready identification of erroneous information which may result from a malfunctioning of the whole system, and if possible, from an incorrect setting.

f. The navigation system must provide in itself maximum practicable protection against the possibility of input blunder, incorrect setting, or misinterpretation of output data.

g. The navigation system must provide adequate means for the pilot to check the accuracy of airborne equipment.

h. The navigation systems must provide information indications which automatically and radically change the character of its indication in case a divergence from accuracy occurs outside safe tolerance.

i. The navigation system signal source element must provide timely and positive indication of malfunction.

j. The navigational information provided by the systems must be free from unresolved ambiguities of operational significance.

k. Any source-referenced element of the total navigation systems shall be capable of providing operationally acceptable navigational information simultaneously and instantaneously to all aircraft which require it within the area of coverage.

l. In conjunction with other flight instruments, the navigation system must in all circumstances provide information to the pilot and aircraft systems for performance of the following functions:

Continuous tracking guidance.
Continuous determination of distance along track.
Continuous determination of position of aircraft.
Position reporting.
Manual or automatic flight.

The information provided by the navigation system must permit the design of indicators and controls which can be directly interpreted or operated by the pilot at his normal station aboard the aircraft.

m. The navigation system must be capable of being integrated into the overall ATC system (communications, surveillance, and navigation).

n. The navigation system should be capable of integration with all phases of flight, including the precision approach and landing system. It should provide for transition from long-range (over water) flight to short-range (domestic) flight with minimum impact on cockpit procedure/displays and workload.

o. The navigation system must permit the pilot to determine the position of the aircraft with an accuracy and frequency that will (a) ensure that the separation minima used can be maintained at all times, (b) execute properly the required holding and approach patterns, and (c) maintain the aircraft within the area allotted to the procedures.

p. The navigation system must permit the establishment and the servicing of any practical defined system of routes for the appropriate phases of flight.

q. The system must have sufficient flexibility to permit changes to be made to the system of routes and placement of holding patterns without imposing unreasonable inconvenience or cost to the providers and the users of the system.

r. The navigation system must be capable of providing the information necessary to permit maximum utilization of airports and airspace.

s. The navigation system must be cost-effective to both the Government and the users.

t. The navigation system must employ equipment to minimize susceptibility to interference from adjacent radio-electronic equipment and shall not cause objectionable interference to any associated or adjacent radio-electronic equipment installation in aircraft or on the ground.

u. The navigation system must be free from signal fades or other propagation anomalies within the operating area.

v. The navigation system avionics must be comprised of the minimum number of elements which are simple enough to meet, economically and practically, the most elementary requirements, yet be capable of meeting, by the addition of suitable elements, the most complex requirements.

w. The navigation system must be capable of furnishing reduced service to aircraft with limited or partially inoperative equipment.

x. The navigation system must be capable of integration with the flight control system of the aircraft to provide automatic tracking.

y. The navigation system must be able to provide indication of a failure or out-of-tolerance condition of the system within 10 seconds of occurrence during a nonprecision approach.

Arguable point

From an applications point of view it becomes quite obvious that GPS can answer all the requirements set forth for a modern aviation-radionavigation system. The mechanical limitations appear to be only within the GPS receiver/processor equipment and interfacing same with the pilot in a productive manner.

6

The FAA and GPS

THE FAA IS CURRENTLY in the process of studying GPS and developing the requirements for its use in the NAS—national airspace system. Started before the GPS system was completed and operational, this study includes defining a set of appropriate standards for GPS aviation receivers and developing the methods for air traffic control handling of GPS aircraft operations (Fig. 6-1 on page 86). There is close cooperation between the FAA, DOD, and the GPS industry in these efforts (Fig. 6-2 on page 87).

A term very important to FAA implementation of GPS into air traffic control is *supplemental navigation*. The primary difference between a supplemental navigation system and a system which meets RNP (required navigation performance) is both the availability of the navigation signal and its integrity.

FAA studies and research

Based upon the studies and research of GPS by the FAA, it has been determined that:

- GPS accuracy of 100 meters (2 drms) is suitable for all current civil aviation accuracy requirements except precision approach and landing.
- The GPS coverage provided by SPS (standard positioning service) has the potential to provide required navigation performance for most phases of flight.
- A U.S. National Aviation Standard for GPS has been prepared to support the supplemental use of GPS.
- Resulting from flight tests, there will be about 5000 instances where properly equipped aircraft can make GPS nonprecision approaches.
- Investigations of GPS/LORAN-C integrated operations and interoperability have been completed.
- GPS user equipment will probably cost more than VOR receivers for general aviation, but will be about the same as other RNAV equipment.

- The current DOD GPS satellite and control segment failure warning system does not provide warnings soon enough after an out-of-tolerance condition occurs to be suitable for civil GPS-approach integrity.
- Operational and performance standards for GPS avionics have been prepared and published in the form of TSO-C129.
- RAIM (receiver autonomous integrity monitoring), using barometric altitude as an input, is being developed to increase GPS integrity for IFR approaches.
- A DGPS (differential GPS) is required to support landing approaches, at lower minimum altitudes than is possible using SPS signals alone.

The FAA is continuing to determine and develop a suitable means of augmenting GPS to meet RNP requirements for the NAS. Currently being examined are:

- The need for continued modernization, maintenance, and sustaining engineering of VOR/DME
- The performance of LORAN-C as a supplement to VOR/DME use and the implementation of LORAN-C as a nonprecision approach aid
- Further need and/or development of the MLS (microwave landing system)

The activities of the FAA are broadly directed toward improving navigation systems serving civilian and military aviation users. These cover five phases of flight:

- Oceanic and domestic enroute
- Nonprecision approach
- Remote area operations
- Helicopter IFR operations
- Precision approach and landing

The FAA navigation program has three specific goals:

1. To provide information that will support FAA recommendations on the future mix of navigation aids.
2. To assist in the near-term integration of new/existing navigation aids into the NAS as supplements to VOR/DME.
3. To provide information that will support the definition of long-term navigation opportunities.

Automatic dependent surveillance

Automatic dependent surveillance is defined as a function in which aircraft automatically transmit navigation data derived from on-board navigation systems via a data link for use by air traffic control. In effect, it is radar without the radar. The aircraft, knowing its position, will send that information to the controller's screen for visual display. The display can be highly augmented with computed information to assist the controller.

Automatic position report processing and analysis will result in nearly real-time monitoring of aircraft movement. Automatic flight plan deviation alerts and conflict (collision avoidance) probes will support reductions in separation minima and increased accommodation of user-preferred routes. Graphic display of aircraft movement and automated processing of data messages, flight plans, and weather information will significantly improve the ability of the controller to interpret and respond to all situations without an increase in workload.

Currently a navigation-based system of automatic aircraft position reporting is being evaluated for application in areas lacking radar surveillance. The system, LORAN-C Flight Following, has been installed in the Houston Air Route Traffic Control Center (ARTCC) and is used to enhance ATC operations in the offshore helicopter sector of the Gulf of Mexico.

GPS IMPLEMENTATION PHASES

The first form of approved instrument landings using GPS came in the form of nonprecision "overlay" approaches on June 9, 1993, with the implementation of Phase I (the first of three implementation phases of GPS navigation approval by the FAA).

Phase I

Phase One allows for the primary use of GPS for IFR operations providing that traditional navigation equipment is:

- On-board
- Operating properly
- Monitored on a supplemental basis

Specifically, it is required that the GPS approach be monitored using the designated ground navaid for that approach.

GPS overlay approaches are authorized for nonprecision VOR, VOR/DME, NDB, NDB/DME, TACAN, RNAV, and special-instrument approaches. Generally, per TSO-C129, GPS approaches must be embedded in the receiver's database and the approach waypoints be displayed in sequential order.

Phase II

Phase Two implementation was announced on February 17, 1994, authorizing GPS as a sole means of navigation providing the GPS equipment meets the criteria of TSO-C129 including RAIM. However, traditional navigation equipment must be on-board and operational, although there is no requirement for monitoring same.

Note: Both Phase I and Phase II are transparent to the air traffic control system.

	1991	1992	1993	1994	1995	1996	1997	1998	1999	2000	2001	2002	2003	2004	2005
Oceanic enroute	multi-sensor		supplemental		RNP										
Domestic enroute	multi-sensor		supplemental							RNP					
Terminal		multi-sensor	supplemental							RNP					
Nonprecision approach		multi-sensor	supplemental							RNP					
Cat I precision approach				special use										RNP	
Cat II/III precision approach				feasibility											

   – GPS as input to multi-sensor navigation

– GPS as supplemental navigation

– GPS augmented to satisfy RNP

– GPS augmented for special use

– Feasibility determined

Fig. 6-1. Projected civil aviation GPS operational implementation chart.

Phase III

Phase Three will occur when DOD declares GPS FOC (full operational capability), which is expected in 1995. This will allow full and sole use of GPS equipment meeting TSO-C129.

This final phase is characterized by changing approach procedure names to include GPS—i.e., "GPS RWY 36"—and deleting the requirement for the ground navaid supporting the approach to be operational. Pilots and ATC controllers will begin specific phraseology for GPS approach requests and clearances during this last phase.

GPS EQUIPMENT REQUIREMENTS

The FAA, through TSO-C129, has classified GPS equipment by intended use and capabilities:

- *Class A*—Equipment incorporating both the GPS sensor and navigation capability (stand-alone GPS receiver/processor as generally found in general aviation). This equipment shall incorporate RAIM (receiver autonomous integrity monitoring) as required.

  ~*Class A1*—Enroute, terminal, nonprecision approach (except localizer, LDA (localizer directional aid), and SDF (simplified directional facility) navigation capability.

  ~Class A2—Enroute and terminal navigation capability only.

	Flight phase														
	Oceanic			Enroute			Terminal			Nonprecision approach			Precison approach		
	A	I	C	A	I	C	A	I	C	A	I	C	A	I	C
GPS (with altimeter)	●	○	?	●	○	?	●	○	?	●	○	?	○	○	?
GPS (with GIC)	●	●	?	●	●	?	●	●	?	●	●	?	○	○	?
GPS/IRS	●	●	●	●	●	●	●	●	?	●	●	?	○	○	?
GPS/Loran-C	●	●	?	●	●	●	●	●	●	●	●	●	○	○	?
GPS/GLONASS	●	●	●	●	●	●	●	●	●	●	●	●	○	○	?
DGPS	F/A			F/A			●	●	?	●	●	?	●*	●	?

*Potential for near CAT 1

Parameter code

A   Accuracy performance
I   Integrity capability
C   Coverage/availability
F/A Possible future application

Performance level

● Adequate

○ Does not have

? Uncertain

Fig. 6-2. Potential performance summary by flight phase augmentation for RNP (required navigation performance).

- *Class B*—Equipment consisting of a GPS sensor that provides data to an integrated navigation system (i.e., flight management system, multi-sensor navigation system, etc.).

    ~Class B1—Enroute, terminal, and nonprecision approach (except localizer, LDA, and SDF) capability. This equipment provides RAIM capability as required.

    ~Class B2—Enroute and terminal capability only. This equipment provides RAIM capability as required.

    ~Class B3—Enroute, terminal, and nonprecision approach (except localizer, LDA, and SDF) capability. This equipment requires the integrated navigation system to provide a level of GPS integrity equivalent to that provided by RAIM.

    ~Class B4—Enroute and terminal capability only. This equipment requires the integrated navigation system to provide a level of GPS integrity equivalent to that provided by RAIM.

- *CLASS C*—Equipment consisting of a GPS sensor that provides data to an integrated navigation system (i.e., flight management system, multi-sensor navigation system, etc.) which provides enhanced guidance to an autopilot or flight director in order to reduce flight technical error. This class of equipment is limited to installations in aircraft approved under FAR (federal aviation regulation) Part 121 or equivalent criteria. (It is intended that this class of equipment need not meet the display requirements applicable to the other equipment classes of this TSO).

  ~Class C1—Enroute, terminal, and nonprecision approach (except localizer, LDA, and SDF) capability. This equipment provides RAIM capability as required.

  ~Class C2—Enroute and terminal capability only. This equipment provides RAIM capability as required.

  ~Class C3—Enroute, terminal, and nonprecision approach (except localizer, LDA, and SDF) capability. This equipment requires the integrated navigation system to provide a level of GPS integrity equivalent to that provided by RAIM.

  ~Class C4—Enroute and terminal capability only. This equipment requires the integrated navigation system to provide a level of GPS integrity equivalent to that provided by RAIM.

Note: Systems utilizing VOR and/or DME for integrity monitoring may require modification in the future as changes to the national airspace system occur.

User interface

TSO-C129 specifies the design and capabilities of the user interface (controls and display):

Controls shall be designed to maximize operational suitability and minimize pilot workload. Reliance on pilot memory for operational procedures shall be minimized.

Controls that are normally adjusted in flight shall be readily accessible and properly labeled as to their function.

The equipment shall be designed so that all displays and controls shall be readable under all normal cockpit conditions and expected ambient light conditions (total darkness to bright reflected sunlight). All displays and controls shall be arranged to facilitate equipment usage.

RAIM implementation

The requirements of RAIM during GPS use are outlined in TSO-C129:

The RAIM function shall provide terminal integrity performance within 30 nm of the departure and destination points. In addition, approach mode (class A1 equipment) integrity performance shall be provided from 2 nm prior to the final approach fix to the missed approach point. Enroute integrity performance shall be provided during other conditions.

The equipment shall automatically select the appropriate RAIM integrity performance requirements.

Equipment certified as class A1 shall:

Upon transition to approach integrity, automatically verify (via RAIM prediction function) that satellite vehicle geometry will be suitable during nonprecision approaches to enable the RAIM function to be available upon arrival at the final approach fix and the missed approach point. Satellite vehicle failures (detected and deselected by the equipment) that occur after the final approach fix that prevent the RAIM detection function do not require annunciation for a period of 5 minutes.

Provide the pilot, upon request, a means to determine if RAIM will be available at the planned destination at the ETA (estimated time of arrival), within at least ±15 minutes computed intervals of 5 minutes or less. Once complete almanac data has been received, this capability shall be available at any time after the destination point and ETA at that point are established. The availability of corrected barometric altitude, either by automatic or manual altimeter setting input, may be assumed for this purpose.

Display, upon request, RAIM availability at the ETA and over an interval of at least ±15 minutes computed in intervals of 5 minutes or less about the ETA.

GPS FOR INSTRUMENT APPROACHES

A key FAA project leading to the use of satellite technology for air navigation was the GPS Approach Overlay Proof of Concept Project. This project demonstrated that existing nonprecision approaches (except those localizer-based) may be safely and efficiently flown using TSO certified GPS avionics.

The FAA states this project was typical of their "user/industry involvement" philosophy. Participants in the flight testing included the Aircraft Owners and Pilots Association (AOPA), Transport Canada, and Volpe Transportation Systems Center.

The overlay concept

The overlay concept was instituted to quickly and inexpensively approve a large number of nonprecision approaches for GPS use. Initially, nearly 5000 nonprecision approaches will become available for GPS use. Overlays allow the FAA to merely place the GPS approach over an existing nonprecision VOR, VOR/DME, NDB, NDB/DME, etc., approach (Figs. 6-3, 6-4, and 6-5). This is a cost-saving means of eliminating the need to develop new approaches and saves airports the expense of certifying a new and specific GPS approach.

Ultimately, 2000 or more airports will be added to the GPS nonprecision approach list. This will mean that nearly 50 percent of the airports in the U.S. will have GPS nonprecision approach procedures.

The GPS approach is made via programmed waypoints. Each step of the approach will be indicated by a waypoint and the approach flown from one waypoint to the next in a sequential manner. The FAA, recognizing the necessity for

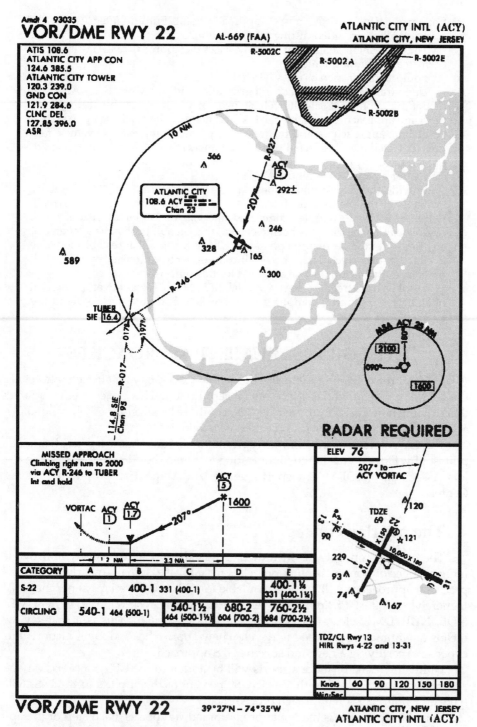

Fig. 6-3. VOR/DME RWY 22 instrument approach procedures for Atlantic City International (ACY). (For illustrative use only—not for navigation use.)

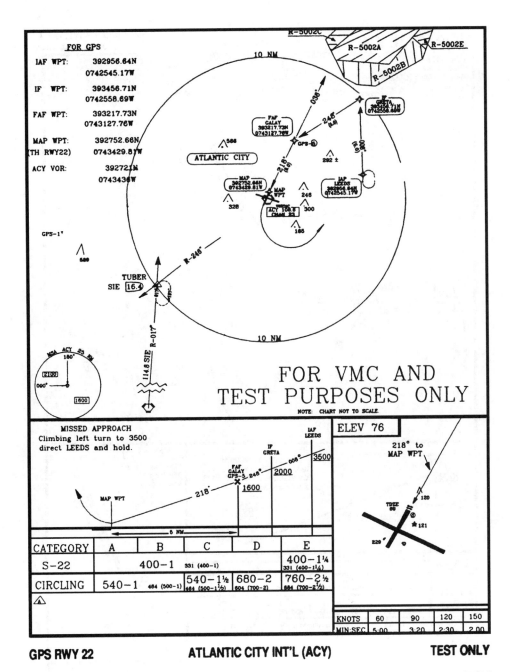

GPS RWY 22　　　**ATLANTIC CITY INT'L (ACY)**　　　**TEST ONLY**

Fig. 6-4. GPS RWY 22 Instrument Approach Procedures for Atlantic City International (ACY) used for nonprecision GPS testing. Notice the use of geographical coordinates to indicate each point of the approach. (For illustrative use only—not for navigation use.)

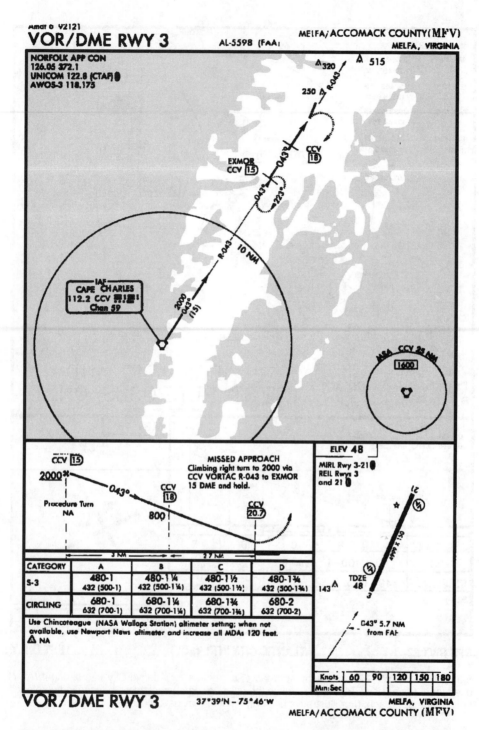

Fig. 6-5. This simple NDB approach for Accomack County Airport is typical of the many that will be overlayed for GPS. (For illustrative use only—not for navigation use.)

accuracy in programming the approach waypoints into the receiver's database, requires that the approaches be canned into the database by an outside and approved source. TSO-C129 requires:

The equipment shall provide an appropriately updatable navigation database containing at least the following location information in terms of latitude and longitude with a resolution of 0.01 minute or better for the area(s) in which IFR operations are to be approved: all airports, VORs (and VORTACs), NDBs, and all named waypoints and intersections shown on enroute and terminal area charts, Standard Instrument Departures (SIDs) and Standard Terminal Arrival Routes (STARs).

Note: Manual entry/update of the navigation database shall not be possible. (This requirement does not preclude the storage of "user defined data" within the equipment.)

Further, the equipment shall store all waypoints, intersections, and/or navigation aids and present them in the correct order for a selected approach as depicted on published nonprecision instrument approach procedure charts. The sequence of waypoints shall consist of as least the following:

Selected IAF (initial approach fix)
Intermediate fix(es)
Final approach fix
Missed approach point
Missed approach holding point

For procedures with multiple IAFs, the system shall present all IAFs and provide the capability for pilot selection of the desired IAF. Selection of the desired IAF shall automatically insert the remaining waypoints in the approach procedure in the proper sequence.

Waypoints utilized as a final approach fix or missed approach point in a nonprecision approach procedure shall be uniquely identified as such to provide proper approach mode operation.

Modification of the data associated with the published instrument approach procedures by the user shall not be possible.

Waypoint data utilized in nonprecision approach procedures shall be in terms of latitude and longitude and cannot be designated in terms of bearing (radial) and distance to/from a reference location.

When in the approach mode, except for holding patterns and procedure turns, the equipment must establish the desired flight path in terms of the path between defined endpoints up to the missed approach point.

The equipment shall provide the capability for entering, storing, and designating as part of the active flight plan a minimum of 9 discrete waypoints (including the active waypoint). In addition, for class A1 equipment, it shall store and designate as part of the active flight plan the complete sequence of waypoints from the navigation database necessary to complete the selected approach including the missed approach.

Waypoints shall be coded in the navigation database to identify them as "fly by" (turn anticipation permitted) or "fly over" (turn anticipation not permitted) as required by the instrument approach procedure, SID, or STAR. Waypoints which define the missed approach point and missed approach holding point in instrument approach procedures shall be coded as "fly over".

Navigation databases shall meet the standards specified under "Preparation, Verification, and Distribution of User Selectable Navigation Databases" and "User Recommendations for Aeronautical Information Services."

The method of GPS navigation the pilot will use in flying a canned GPS nonprecision approach is *TO-TO*. In other words TO a waypoint and then proceed TO the next waypoint in sequential order. As called for by TSO-C129:

> The equipment shall provide the capability to fly from the present position direct to any designated waypoint. Access to this feature shall be by means of a single action by the pilot. Selection of the desired "TO" waypoint may require additional actions.

Category I approaches

After approval of nonprecision GPS approaches in 1993 the FAA has devoted its attention to implementing a precision approach capability. The primary problem is the lack of accuracy sufficient to support precision approaches.

Initially, the FAA issued Special-Use approval to specific users at specified locations for an aviation user to fly a particular approach. The organization or user purchased the required ground and airborne equipment and would petition the FAA for a Special Use approval. This program was operated strictly on a case by case basis and provided specific application testing at no hardware expense to the government.

An example of a Special Use is the Juneau, Alaska, Project. This cooperative effort, currently underway, has the goal of providing a precision approach capability for Alaska Airlines (the aviation user) into the Juneau airport (specific approach). In addition to Alaska Airlines, other participants include the FAA Satellite Program Office, the FAA Alaskan Region, Litton Avionics, Trimble Navigation, and officials from the State of Alaska and City of Juneau.

The existing weather minimums for landing at Juneau are 1000 foot ceiling and 2 miles of visibility. Numerous times during the year the weather does not permit landings or departures.

Because Juneau is the state capital of Alaska and is accessible only by air or sea (there are no roads into the city), it would be extremely beneficial to improve the landing capability at the airport. The present ILS systems are not feasible due to terrain restrictions (mountains). It is believed that GPS with a differential capability, can help reduce the existing weather minimums.

Special-Use programs will eventually lead to acceptance of DGPS for precision approaches and will particularly shine in difficult instances where conventional straight approaches are not possible. Additionally, through use studies such as these, a determination can be made as to the most appropriate form of DGPS to use for aviation purposes.

Category II and III approaches

Although Category I approaches appear to be technically feasible and practical, Category II and III precision approaches are more challenging because of their greater accuracy and integrity demands. Originally, the FAA viewed this as a long-term technical problem.

Recently, however, industry research into GPS-augmentation techniques such as DGPS and kinematic carrier phase tracking (means of determining PVT by radio signal carrier phase matching rather than PRN code correlation) has produced promising results. Consequently, the FAA is accelerating its research program to examine the technical feasibility of satellite-based Category II/III approaches.

The FAA's feasibility study of CAT II/III approaches will be completed by 1995. This will enable the FAA and other branches of the United States government to make informed judgments about the future precision approach replacement for ILS including MLS decisions.

The focus of this near-term effort is not to develop a GPS CAT II/III precision approach system, but rather to enable FAA decision makers to arrive at an informed decision about future precision approach aid procurement for the FAA. Any satellite-based system must also prove to be producible and economical. It must also readily integrate into the national airspace system.

DGPS

As mentioned previously, the FAA is actively pursuing technology related to the use of differential corrections to the standard positioning service (SPS) of GPS. This includes the examination of both LADGPS (local area differential GPS) and WADGPS (wide area differential GPS).

The purpose for examining DGPS is to enable the FAA to solve questions concerning accuracy, integrity, and availability so that the SPS portion of GPS can be utilized for all phases of flight and particularly for precision approaches to landing.

The FAA is actively supporting the activities of the International Civil Aviation Organization and the Radio Technical Commission for Aeronautics task forces in the definition of the Global Navigation Satellite System and the associated implementation planning guidelines. These efforts will assure that satellite-navigation capabilities are implemented in a timely and evolutionary manner on a global basis.

Integrity algorithms have been defined, developed and are being tested. The primary goal of the integrity development is to accurately assess the GPS-satellite health and warn pilots of failed satellites within six seconds to support the ILS CAT I approach requirements.

The current focus is evaluation of the GIB (GPS Integrity Broadcast) and WADGPS. GIB testing addresses integrity issues such as:

- time to detect
- false alarm rates
- data formats
- detection algorithms

The WADGPS testing is examining the potential accuracy over long baselines, defining spatially-decorrelating errors and modeling same, and the networking of remote and master stations.

The primary advantage of a wide-area system is the reduction of equipment costs. The primary disadvantage of the system is the reduction of accuracy due to spatially-decorrelating errors. To mitigate these errors, atmospheric modeling and pin-point locational data have been developed and will be applied.

Preliminary flight-test data revealed wide-area results that were better than expected. The initial flight testing began early in 1993 and incorporated a VHF data link while simulating the delay associated with a satellite link. Later flight testing will use an INMARSAT geostationary satellite link for transmission of the DGPS signal.

GPS INSTALLATIONS

There are two basic GPS installations: the *Stand-alone* GPS system that is not integrated with any other navigation system to derive its position, and the *Multiple-sensor* navigation system that may include a GPS sensor with one or more FAA-approvable navigation sensors, all of which operate independently (not integrated).

Multiple sensor systems do not provide a blended position solution or position integrity comparison. In essence they are separate entities housed in a single box. Therefore, an IFR-approved non-GPS sensor can retain its approval for use under IFR conditions when a GPS sensor is collocated with it.

The installation of GPS equipment used as a stand-alone system or as part of a multiple-sensor navigation system, as defined above, can be approved using the following considerations:

For VFR GPS installations—Operators wishing to use GPS for operations limited to VFR—may obtain approval of the installation by TC (type certification), STC (supplemental type certification), the FAA field approval process, or by use of previously approved data.

The logbook approval for return to service must be signed by one of the entities noted in FAR 43—i.e., repair station, manufacturer, holder of an inspection authorization, etc.). The installation verification should ensure, at a minimum:

- That the GPS installation does not interfere with normal operation of other equipment installed in the aircraft. Verify by a ground test and flight test to check that the GPS equipment is not a source of objectionable EMI (electromagnetic interference), is functioning properly and safely, and operates in accordance with the manufacturer's specifications.

- That the structural mounting of the GPS equipment is sufficient to ensure the restraint of the equipment when subjected to the emergency landing loads appropriate to the aircraft category.

- That a navigation source annunciator is provided on or adjacent to the display if the GPS installation supplies any information to displays, such as a HSI (horizontal situation indicator) or CDI (course deviation indicator), which can also display information from other systems normally used for aircraft navigation.

- That the GPS controls and displays are installed with a placard(s) which states "GPS Not Approved for IFR."

- That the GPS may be coupled to the "radio nav" function of an autopilot provided the system has a CDI or steering output that is compatible with the autopilot.

- That the outputs from a nonintegrated GPS receiver providing any information to displays, meaning CDI, HSI, etc., must be designed using accepted aeronautical practices, perform their intended function, and have no complex switching or operational features. Such installations may use the limitations and normal or emergency procedures supplied by the system manufacturer for the end user. Installations that require complex switching procedures or have functions that may result in information or maneuvers that are misleading or unacceptable must have a flight manual supplement, or a supplemental flight manual for aircraft without an FAA approved flight manual, that includes any limitations or cautions and operating procedures.

Multi-sensor navigation system approval

The FAA says that a multi-sensor navigation system incorporating a GPS sensor may be approved for IFR or VFR use provided:

> The airworthiness considerations contained in AC 20–130 or equivalent (for use in the U.S. National Airspace System (NAS) and Alaska) and AC 12033 (for operation in the North Atlantic Minimum Navigational Performance Specifications (MNPS) airspace and other oceanic or remote areas), if applicable, are followed.
> A flight manual supplement (or supplemental flight manual for aircraft without a FAA approved flight manual) is required that includes any limitations, operating instructions, and the following caution:
> Except as specified by this flight manual, the GPS satellite constellation may not meet the coverage, availability, and integrity requirements for civil aircraft navigation equipment. Users are cautioned that satellite availability and accuracy are subject to change, and appropriate GPS status information should be consulted.

Position information must be available, at all times, from at least one other approved or approvable sensor, appropriate for geographic area and flight phase, for IFR operations. The multi-sensor navigation system must monitor the integrity of the GPS information by comparing the difference between the position computed using GPS information and the position computed from the other approved sensor(s). Although a system may provide the necessary level of integrity in various ways, such as using comparisons related to a system mode or configuration rather than flight condition (phase of flight), the difference between GPS and other sensor positions should not exceed the following values (unless approved by the FAA):

Flight condition	Monitor limit
Oceanic or Remote Areas	12.6 nm
Enroute IFR along random routes	3.8 nm
Enroute IFR on airways in the NAS	2.8 nm
Terminal IFR operations in the NAS	1.7 nm
Instrument approach operations in the NAS	0.3 nm

The system must detect when any sensor cannot provide the accuracy required or is not available. An advisory indication must be provided to the flight crew.

The system must detect when sensors (other than GPS) required for enroute, terminal, or nonprecision approach operations are not of the required accuracy or not available. Under such conditions, the flight crew must be alerted that the system does not meet IFR requirements.

In the approach mode, the system must detect when sensors (other than GPS) are not of the required accuracy or not available. Under such conditions, a failure indication must be displayed on the dedicated navigational display.

For IFR multi-sensor systems, the minimum performance standards specified in TSO-C115a or an acceptable alternate means must be met.

The multi-sensor approval option requires a type certificate or supplemental type certificate for the initial approval. Follow-on installation approvals may be accomplished by TC or STC or may be in the form of a field approval on an FAA Form 337 provided the data initially approved is applicable to the follow-on approval. The applicant or installing entity accomplishing a follow-on multi-sensor system installation utilizing this field approval method should follow the procedure in AC 20–130, para. 9.b.(1)–(6).

Follow-on installations using the field approval process should use the sample aircraft flight manual supplement in appendix 2 of AC 20–130 based on the data from the initial approval. The FAA inspector signing the supplement should consider the note on page 3 of appendix 1 in AC 20–130.

Note: Some materials contained in this chapter are excerpted from FAA documents.

7

GPS receivers

RECEIVING EQUIPMENT FOR GPS, as envisioned by the typical user, consists of a combined set of components working together to provide the user with PVT (position, velocity, time) information in a usable form. Customarily this means a single unit containing a receiver, data processor, and a control/display panel (user interface).

GPS equipment intended for general aviation use is usually capable of providing additional information, such as: estimated times enroute and of arrival, distance and bearing to/from specified locations, speed calculations, user positions generally referred to as waypoints, and a database of aviation facilities.

Phraseology

In the GPS industry there are several phrases used to describe such a complete device:

GPS receiver—although technically indicating only a single component in the user device, marketing and sales forces have bent the definition to mean a complete device.

Receiver/processor—a complete unit capable of determining PVT and other information. Specifically, it indicates a combination of a receiver and connected data processor.

User equipment (UE)—is a term found in many government and military documents meaning the total equipment designed to fulfill a specified GPS user requirement.

Generally, all of the above terms can be used interchangeably for the end-user's purposes, describing a complete device from antenna to readout display.

Generic GPS receiver hardware

The basic GPS receiver, built of a generic architecture shared by all GPS receivers (Fig. 7-1), requires:

- L-band antenna to receive the satellite transmitted signal
- PM (phase modulated) radio receiver
- Data processor for calculations
- User interface (control panel and visual output)

Simply stated: the antenna receives the GPS signals, the receiver determines the PRN ranging codes on one or both of the L-band RF carrier waves, generates PR measurements, and demodulates the 50 Hz navigation message data, and the data processor solves the four-equation/four-unknown positioning equations. The device communicates with the user through an interface which typically consists of a display and assorted controls.

GPS receiver operation

The GPS receiver receives and tracks satellite signals to make pseudorange and deltarange (velocity) measurements. It must also collect NAV-msg data as a reference source.

To make a satellite selection, the receiver is required to have a basic knowledge of the satellites in orbit. This knowledge, called *almanac information*, includes what satellites are operating normally and are healthy, the available satellite PRN codes, and where the satellites are orbitally located (geometry). The receiver gets this information from the almanac data contained in subframes 4 and 5 of the NAV-msg.

Once the almanac information has been obtained, the receiver uses knowledge about its current position and time (either stored in its internal memory or input by the user) to determine which satellites should be visible in the sky overhead and what their relative geometry is. The receiver then selects the best set of four satellites to track.

An operating assumption for all GPS receivers is the selection of the best set of four satellites to use for position/time solution from all those visible at a given time and location. Only those above the masking area of the local horizon, normally considered to be 5 degrees for precise positioning service and 7.5 degrees for standard positioning service, are considered.

The purpose for the masking area near the horizon is to provide greater accuracy. Satellite-signal strength is not a factor, as power levels are adequate for reception down to the horizon. However, at low-elevation angles (angle between the satellite and the horizon) the signal is subject to elevated path bending and delays caused by atmospheric distortions including:

1. The tropospheric path bending and
2. ionospheric path delays.

Receivers having two frequencies can compensate for these atmospheric distortions based upon a comparison between the two received signals, however, for general-aviation purposes a single frequency receiver is the norm (remember, SPS is what civilians use). Although both path aberrations can be somewhat mathematically corrected, using models or forecasts, the accuracy of the tropospheric

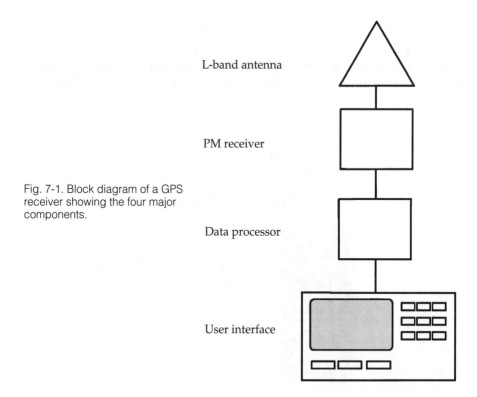

L-band antenna

PM receiver

Fig. 7-1. Block diagram of a GPS receiver showing the four major components.

Data processor

User interface

model and ionospheric model degrade as the signal passes through more of the atmosphere on its way to the receiver.

Restricting the tracking of satellites to those above some minimum elevation angle (masking area) prevents an accumulation of compensation errors in its PR measurements. The mathematical models are most accurate when the satellites are directly overhead and least accurate when they are at low elevation angles.

Best four satellites

For the purposes of aviation the "best set" of four satellites indicates a combination of those satellites available for use, giving the lowest GDOP (geometric dilution of precision) value. This combination will optimize 3D (three-dimensional) PVT (position, velocity, and time) accuracy. Note that for marine operations the best four satellites are selected for the lowest HDOP (horizontal dilution of precision), as vertical position and time accuracy are less important in seaborne navigation than in other applications.

The NAV-msg contains satellite health information for all the satellites in the GPS constellation. Each satellite broadcasts health summaries for all the other GPS satellites in subframes 4 and 5 (almanac data). The NAV-msg also contains health information about the broadcasting satellite in subframe 1.

Note: The health data in subframes 4 and 5 are updated less frequently than subframe 1. Subframe 1 may be used to indicate short-term health problems or may be updated before subframes 4 and 5. A satellite should never be used if its NAV-msg indicates it to be unhealthy.

As long as there are more than four healthy satellites visible, there will be many different combinations available for the receiver's use. The receiver will automatically select which satellites to track with no user input (Fig. 7-2).

Fig. 7-2. The bargraph display indicates the relative signal strengths of the satellites available.

Garmin International

GPS receiver start-up

The TTFF (time to first fix) is the elapsed time necessary for the GPS receiver to acquire satellite signals, navigation data, and determine the initial position solution (Fig. 7-3). It includes cold start-up, which means from the time a receiver is first turned on, including warm-up and a frequency stabilization period of as much as six minutes.

The first time a receiver is operated it must search for a signal and the included almanac data. After the initial satellite signal is acquired, the almanac data is obtained from the NAV-msg. It can take up to 12.5 minutes to collect the full almanac after initial acquisition. Almanac data can be obtained from any GPS satellite. Proper design of the receiver provides for the storage of almanac data for later use and reduced future start-up time.

After the desired satellites are selected, the ephemeris data, contained in the NAV-msg, must be collected for each satellite received. It can take from 30 seconds to 3 minutes for ephemeris data collection depending upon the receiver's design.

TYPES OF GPS RECEIVERS

There are several types of basic GPS receivers. They vary in complexity, capabilities of tracking the satellites, speed of output (information updating), and application. Further, corporate marketing has clouded the identity of some receiver types

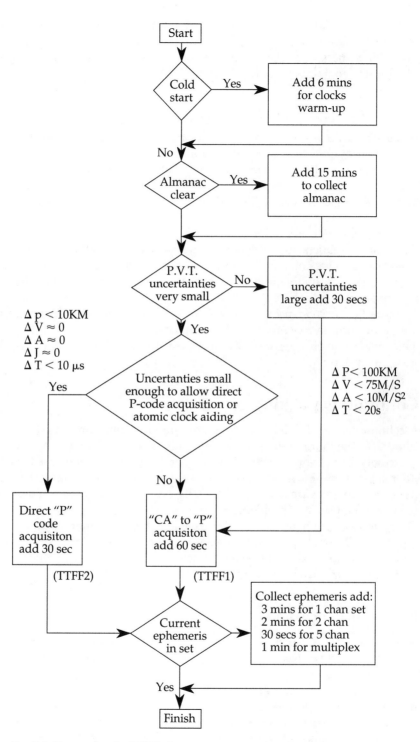

Fig. 7-3. Time to first fix (TTFF) chart. U.S. Government publication

currently on the market. Generally, as the complexities increase so do the capabilities, however, rarely are complexities a concern to the user. One of the greatest features of GPS is the sheer ease of receiver operation from the user standpoint, regardless of the complexities or capabilities.

Sequencing receivers

Sequencing-GPS receivers track satellites by use of one or two hardware (RF) channels. Their operation is a simple stepping from one selected satellite to another on a timed basis for measurements and combining same when all steps have been made.

In general, single channel sequencing receivers are limited to low-dynamic or stationary applications. The reason for this is the distance traveled between steps, assuming the receiver is in motion.

Two-channel sequencing receivers are an improvement as they split receiving duties and can provide enhanced accuracy over single-channel receivers. They are considered adequate for medium dynamic (speed) operations (helicopters and slower airplanes).

An advantage of sequencing receivers is their simple circuitry, low production costs, and low power consumption. Note that recent fast-sequencing single RF channel digital receivers appear to have overcome the drawbacks of traditional sequencing receivers.

Continuous tracking receivers

A continuous-tracking receiver has a minimum of four hardware (RF) channels and tracks the four required satellites simultaneously. These receivers are not limited to only four channels, many have five or more. In the latter cases, the fifth (or other subsequent) channel is used to read the NAV-msg of the next satellite when selection changes are made. Additionally, other satellites are simultaneously tracked to eliminate GDOP problems.

Continuous-tracking receivers offer excellent high-dynamic capabilities so they operate well from fast moving airplanes such as military fighters. They track four satellites continuously—at the same time—thereby providing excellent 3D accuracy. Further, they provide low time to first fix.

Multiplex receivers

A multiplex (MUX) receiver switches at a fast rate between satellites being tracked, continuously collecting data and reading the NAV-msgs. Switching is typically 50 times a second.

The single-hardware (RF) channel multiplex receiver is based upon time sharing and only a single code generator and carrier synthesizer is required to track the satellites. MUX receivers display a carrier to noise ratio 4 to 8 db below (less sensitivity) that of a continuous tracking receiver as a factor of the high speed switching. Multiplexing receivers may be thought of as very fast sequencing receivers.

All-in-view receivers

Normally, GPS receivers select four satellites to perform a 3D position fix and maintain same. However, in instances when a satellite is temporarily obscured from the antenna's view, additional satellite signals must be acquired to maintain continuity of the PVT solution.

An all-in-view receiver tracking more than the required four satellites can instantly replace any lost signal with another already being monitored. There is no resulting loss in continuity. It is not uncommon to see receivers capable of tracking up to eight, or more, satellites.

All-in-view receivers can be multiple channel—one hardware (RF) channel per satellite—or a multiplex scheme. The latter being the least expensive, for hardware.

Digital receivers

Most early GPS receivers were of analog design, whereas the later models are of digital design. The latter allows greater flexibility, reduced manufacturing and maintenance costs, and enhanced capabilities over the early designs (Figs. 7-4 and 7-5).

Early GPS receivers required a dedicated receiver front end for each (hardware) channel used, however, using A/D (analog to digital) conversion techniques a digital multi-channel receiver uses only a single receiver IF (intermediate frequency) for signal amplification and conversion, with signal processing accomplished digitally—making equipment smaller, lighter, and less expensive to build (Fig. 7-6).

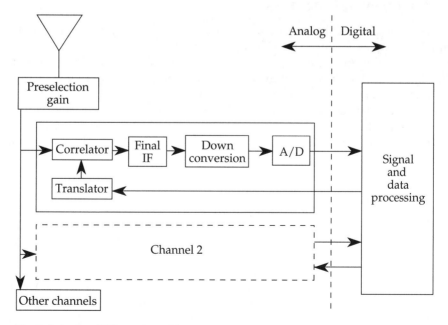

Fig. 7-4. Analog GPS receiver diagram. U.S. Government publication

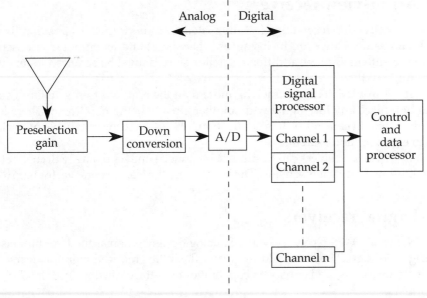

Fig. 7-5. Digital GPS receiver diagram. U.S. Government publication

In essence, a current five-channel digital GPS receiver can be visualized as a single-channel radio receiver with five digital channels, each monitoring an individual satellite.

Differential receivers

As an accuracy improvement and adjunct to GPS receivers, the use of differential techniques requires a specialized type of receiver.

Usually the differential receiver will plug directly into a GPS receiver (Fig. 7-7), if indicated as differential ready, and allow the data processor to include differential corrections when solving for position, velocity, and time.

The receivers may operate in the LF bands for USCG marine-differential signals, FM broadcast band for subcarrier signals (Fig. 7-8), or the VHF band for immediate area signals. At the time of this writing there appears to be no determined standard for aviation use.

User input interface

Some GPS receivers provide for data entry with a keyboard, much like a small calculator. Other units make selections by use of buttons and/or knobs—press a key while numbers or letters scroll by—or by twisting a set of knobs in the manner you dial in a frequency on your nav/com.

The keyboard is quicker and easier to use, although it may be somewhat difficult to operate during turbulent conditions. A full keyboard takes up more physical space than buttons and knobs, and can present mechanical problems, therefore

Fig. 7-6. Although operationally complex, these five-channel GPS receivers are only 3×7 inches in size.

Fig. 7-7. Differential receiver used as an adjunct to a GPS receiver.

Magnavox Electronics Systems

is not found on some compact GPS receivers or most panel-mounted equipment. An interesting fact common to all user interfaces (controls and display) is they use more physical space and have more moving parts than the receiver/processor portion of the overall unit.

Note that the complexity or simplicity of the keyboard and/or display cannot generally be used as a measure of unit capability (Figs. 7-9 and 7-10).

WAYPOINT HANDLING

The most efficient use of GPS for navigation is by employing geographical locations called waypoints. Waypoints are merely locations in standard latitude/longitude format that are either entered manually by the user or carried in a preinstalled database, or a combination of both.

Waypoint storage

Waypoint storage is the receiver's memorization of those locations entered by the operator. Some receivers are limited to a numeric system of waypoint labeling,

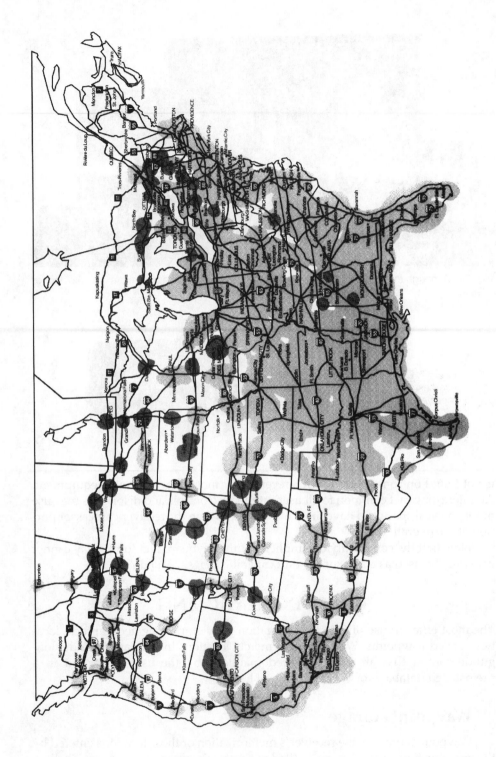

Fig. 7-8. Map of nationwide DGPS service available on FM broadcast subcarriers. OACCOPOINT

Fig. 7-9. A marine-user interface consisting of a complex keyboard and moving map display.

Magellan Systems

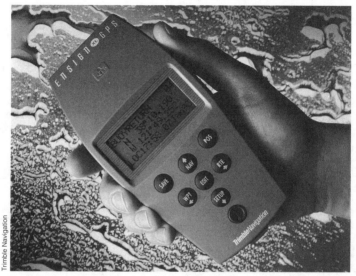

Trimble Navigation

Fig. 7-10. The simple appearance of this GPS receiver is no indication of limited capabilities.

which means that you may have to create a directory to help you remember the waypoint identities, while others allow the selection of more easily remembered names of three to five letters each.

As for storage, some receivers file the waypoints in the order they were entered. More-sophisticated receivers, however, organize the waypoints in alphanumeric order, making them a lot easier and faster, thus more efficient, to retrieve.

Defining a waypoint

The most common means of defining a waypoint is by using its latitude and longitude coordinates. For example, to enter your home airport as a waypoint:

Determine the latitude and longitude of the airport from a source such as the Airport/Facilities Directory or from the meridians and parallels on a VFR Sectional.

Determining from a Sectional is not very accurate due to user-interpretation inaccuracies. Enter same.

Of course you could walk to the middle of the airport and press the "load present position" button, causing the receiver to memorize your present location.

Some units have provisions for entering waypoint data in radial/distance form from a known position. This may simplify your chart-work as it is an accepted concept used in VOR navigation (DME/VORTAC, etc.).

General-aviation databases

A database is a preprogrammed set of information a GPS receiver can use as waypoints. The general-aviation type database will always include geographical coordinates (latitude and longitude) for each point. The points are usually aviation facilities including airports, VORs, TACANs, and NDBs. Other data such as frequencies, airport information, or services available may be included .

Aviation databases may or may not be easily updatable by users. Some are contained in fixed internal memory while others are carried on removable data cards (Fig. 7-11). A few receivers provide for at-home PC interconnection and updating, while still others must be sent to the factory for update service.

Note: TSO-C129 requires special databases for use on nonprecision approaches and that these will not be modifiable by the user.

GPS COMBINED RADIONAVIGATION SYSTEMS

Individual radionavigation systems are sometimes used in combination with other systems. The object of combining systems is to supplement a weakness of another. For example, the combination of a system of high accuracy and low fix rate with a system of lower accuracy and higher fix rate would result in an overall combined system with both high accuracy and high fix rate.

Fig. 7-11. User-insertable databases make updating easy.

Integrated navigation receivers

Integrated navigation receivers combine the signals from multiple sensors to determine position and velocity. The typical sensors of an integrated receiver would include one or more radionavigation receivers and compass/speed sensors.

Recently, several receivers have been developed combining GPS with other radionavigation systems to take advantage of the nearly continuous GPS coverage available as the constellation matures. For the purpose of general aviation these combinations take the form of GPS/LORAN units.

Improved navigation performance is attainable when the time references of different radionavigation systems are related to one another in a known manner and user equipment can more advantageously combine the LOPs (lines of position) from the different systems. Such a relational system is said to be interoperable.

The FAA plans to determine the technical feasibility of combining both GPS and GLONASS (Russian counterpart of GPS) signals in the same user equipment to determine position and for use in navigation. Using information from both these systems would possibly provide more continuous, worldwide coverage than using either system separately—a benefit especially valuable in aviation. At least one manufacturer is independently developing a GPS/GLONASS receiver.

Differential GPS

Although DGPS is not truly a combination of different technologies, it is a means of enhancing the immediate accuracy of positioning (navigation) information as initially explained in chapter 2.

The FAA is considering three types of differential GPS service for aviation use, all three of which are under current research and development:

Local-area DGPS (LADGPS)—located at each airport or closely grouped airports to support instrument approaches down to current CAT I weather minimums.
Wide-area DGPS (WADGPS)—to provide GPS integrity broadcast (GIB) and accuracy improvements for all of North America.
Use of kinematic carrier phase positioning—for instrument approach and landing.

Note: At the time of this writing WADGPS/GIB is in the FAA budget for procurement and installation.

The basic concept of WADGPS/GIB is to have several GPS ground-monitoring stations placed around the country (about 20 for North America). They would be operated through two master control stations where differential corrections and integrity for each satellite are determined.

8

Airplane navigation
with GPS

AIRPLANE NAVIGATION WITH GPS, although extremely accurate, is nearly as simple as point and shoot. However, as simple as it is in use, the pilot must understand some navigational basics about the geographical coordinate system.

Initially, most general-aviation pilots learn to think in terms of radial and distance when navigating with VOR-based equipment. To use GPS, pilots must learn to think in terms of latitude and longitude map coordinates, as they are the position 2D (two-dimensional) method that GPS uses to interface itself with the user. The third dimension, altitude, is used during 3D (three-dimensional) positioning.

Latitude

Latitude lines, called *parallels*, run east and west, parallel to each other starting at the equator. Latitude lines are numbered in angular degrees—0 degrees to 90 degrees north and south of the equator. Only one latitude line circumvents the entire globe, that of 0 degrees, which is the equator. All others become shorter in length as they near the polar areas (Fig. 8-1).

Longitude

Longitude lines, called *meridians* run north and south, connecting the north pole to the south pole and, like parallels, are also spread out angularly. They are numbered in angular degrees from 0 degree to 180 degrees east and 180 degrees west of the Prime Meridian. The Prime Meridian runs north and south through Greenwich, England (Fig. 8-2).

Each line of longitude forms a circle encompassing the entire globe and outlines a plane cutting through the same. All lines of longitude are theoretically the same overall length.

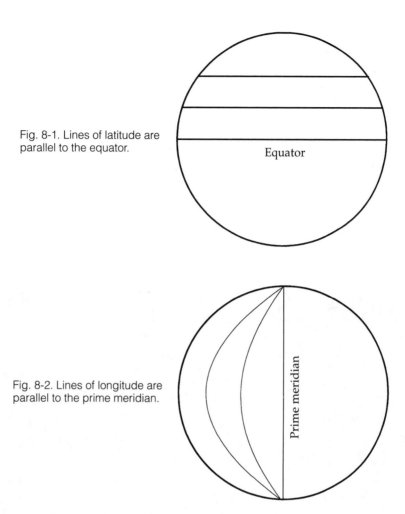

Fig. 8-1. Lines of latitude are parallel to the equator.

Fig. 8-2. Lines of longitude are parallel to the prime meridian.

Map coordinates

Each latitude and longitude degree is divided into 60 minutes and each minute is divided into 60 seconds. It is the map-user's choice whether to use seconds or a decimal fraction of minutes. Seconds are not commonly used in aviation navigation and are replaced with tenths of a minute, except when reading charts. For example:

71 degrees 14 minutes 13 seconds is the same as 71 degrees 14.2 minutes (or if you desire increased precision, 71 degrees 14.217 minutes). Latitude and longitude coordinates can be determined by reference to, for general-aviation purposes, WAC, Sectional, or Terminal-Area National Ocean Service charts.

WAC charts display the lines of latitude and longitude spaced one degree apart, with incremental marks indicating each minute between degrees: 60 minutes per degree.

Sectional charts depict latitude and longitude in a similar manner, except the lines of latitude and longitude are drawn at intervals of a half degree—30 minutes (Fig. 8-3).

Terminal-Area charts show latitude and longitude lines at one quarter degree spacing—15 minutes.

Map reading example: Using the San Antonio Sectional chart you can determine the geographical coordinates of Cole Ranch airport, located north-west of San Angelo, Texas, as 31 degrees 39 minutes north latitude by 100 degrees 58 minutes west longitude (Fig. 8-4).

GPS, however, provides greater precision than the charts, as a matter of interpretation (reading accuracy). Position information is given in terms of degrees, minutes, and hundredths of a minute, and sometimes thousandths. For example:

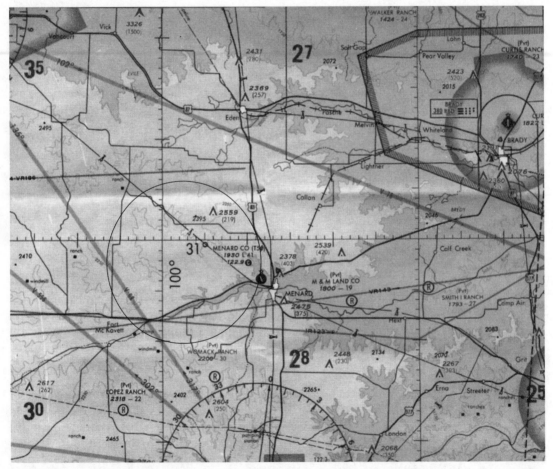

Fig. 8-3. Sectional chart showing latitude/longitude (31 degrees north/100 degrees west). (For illustrative use only—not for navigation use.)

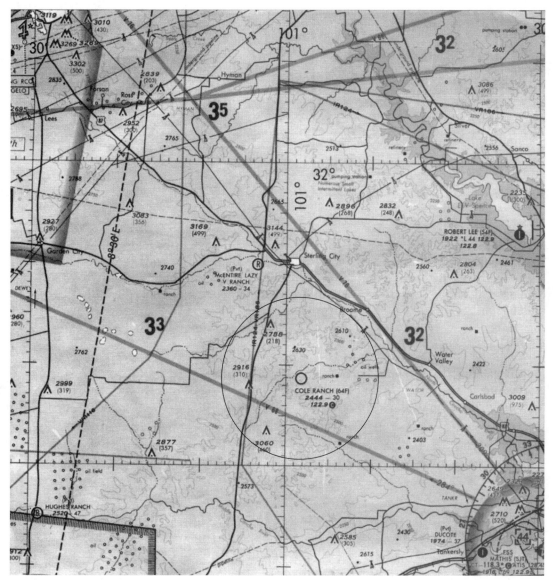

Fig. 8-4. Cole Ranch 31 degrees 39' north /100 degrees 58' west as located on the San Antonio sectional. (For illustrative use only—not for navigation use.)

Cole Ranch airport is 31 degrees 38.62 minutes north by 100 degrees 58.17 minutes west (Fig. 8-5).

Each degree of latitude is equal to 60 nautical miles, therefore a minute equals one nautical mile. A tenth of a minute is approximately 600 feet and a hundredth of a minute is slightly over 60 feet. Geographical references given in seconds can easily be converted to hundredths of a minute by dividing the number of seconds by 60.

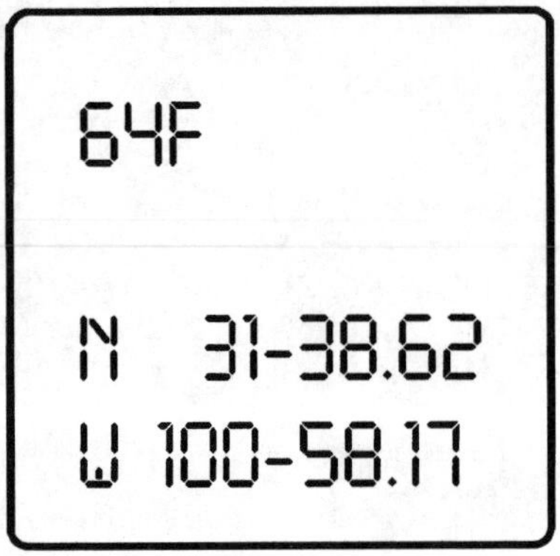

64F

N 31-38.62

W 100-58.17

Fig. 8-5. GPS display showing the location of 64F (Cole Ranch) at 31 degrees 38.62 minutes north /100 degrees 58.17 minutes west. (For illustrative use only—not for navigation use.)

Note: Latitude remains constant at 60 nm per degree. This is not true for longitude. The lines of longitude converge the more pole-ward you travel, causing the spacing between each degree of longitude to decrease. The single exception to this rule occurs exactly at the equator, where each degree of longitude represents the standard 60 nm.

The ups and downs

Considering that most readers of this book will be using GPS in the area of the North-American continent, a simple method of remembering how the numbers of latitude and longitude progress is:

East is least and west is best. The numbers of longitude decrease as you travel to the east and increase as you travel to the west. North is up on your charts and the latitude numbers rise as you go up north. Up is up and down is down.

In the United States and Canada, latitude and longitude are always expressed as *north* and *west* respectively, as these areas are north of the equator and west of the prime meridian.

Why coordinates

Latitude and longitude are the language of the GPS receiver. The geographical-coordinate system was selected for this purpose due to universal applications worldwide. Note, however, there may be some minor discrepancies between specific mapping systems, called datums. For reference to the various datum and co-ordinate systems in use, see appendix C. All the GPS receivers covered in this book use the WGS-84—World Geodetic Survey-1984—datum as a default standard. Other datums are generally user selectable.

Everything calculated by the receiver/processor is based upon geographical coordinates. By comparing current-position coordinates with other-position coordinates, such as waypoints, determinations of speed, times, direction of travel, etc. may be made.

THE PORTABLE GPS RECEIVERS

The gadgetry of today is most impressive in portable form. Years ago there was the hand-held calculator, that indispensable box from the 70s everyone needed. At home, everyone is familiar with full-featured TV remote controls and at the office the laptop computer is king. Now comes GPS.

The portable, or hand-held, GPS receiver is one of the truly remarkable small devices ever made. From a single object we can navigate worldwide with an unbelievable level of accuracy.

Portable GPS receivers have many uses other than aviation. Only a few models are built specifically with aviation in mind. In practice, early aviation portable GPS receivers were simply a successful generic GPS used for surface navigation with an aviation related database installed.

Note: The use of an aviation-version GPS portable receiver is by no means a firm requirement for VFR flying.

Plain portable GPS receivers

For those pilots using GPS for VFR purposes only, an inexpensive means of boarding the GPS ride is the use of a non-aviation hand-held GPS receiver that can have a large number of waypoints entered into it. Non-aviation, for this purpose, means a GPS receiver without an installed aviation related database.

Most GPS receivers have room in memory for 100 or more waypoints, some accept a thousand or more. Because many weekend and evening pilots stay within a couple of hundred miles of their home airport, such a receiver's waypoint storage would be adequate. Waypoints entered might include:

- The home airport
- Favorite destinations
- Navaids generally used

Be sure to leave a few positions open for those extra trips or vacations to places you normally do not go to.

Waypoint sources

Some, but not all, GPS receivers have provisions for entering information in addition to geographical coordinates and assigned names. You might have space to add town names, communication-radio frequencies, runway lengths, etc. No matter what you are entering, you must have accurate sources of the information. Typical reliable sources include:

AIRPORT/FACILITIES DIRECTORY—a government publication updated every eight weeks (Fig. 8-6). It is available by area of the country—Northeast, Southeast, East Central, North Central, South Central, Northwest, and Southwest—and gives pertinent information about all airports, seaplane bases, heliports (open to the public) including: communications data, navigational facilities, and certain special notices and procedures. Airport/Facilities Directories are available from:

TEXAS

COLORADO CITY (TX50) 4 W   UTC – 6(– 5DT)   N32°28.27' W100°55.24'   DALLAS–FT. WORTH
2214   B   FUEL 100LL   H–2E, 5A, L–13A, 15C
RWY 17-35: H5420X60 (ASPH)   S–50   LIRL
AIRPORT REMARKS: Attended irregularly. For fuel call 915–728–2542/2610.
COMMUNICATIONS: CTAF 122.9
   FORT WORTH FSS (FTW) TF 1–800–WX–BRIEF. NOTAM FILE FTW.
RADIO AIDS TO NAVIGATION: NOTAM FILE SJT.
   BIG SPRING (L) VORTACW 114.3   BGS   Chan 90   N32°23.14' W101°29.02'   069° 29.1 NM to fld. 2670/11E.

COLUMBUS
ROBERT R. WELLS JR.   (66R)   3 S   UTC – 6(– 5DT)   N29°38.46' W96°30.95'   HOUSTON
242   B   FUEL 100LL   L–17A
RWY 15-33: H4000X45 (ASPH)   S–12.5   LIRL
RWY 15: Thld dsplcd 654'. Brush.   RWY 33: Brush.
AIRPORT REMARKS: Unattended. Rwy 15–33 S 54' of rwy unlgtd. Rwy 15–33 marked with 6" centerline stripes only.
COMMUNICATIONS: CTAF 122.9
   MONTGOMERY COUNTY FSS (CXO) TF 1–800–WX–BRIEF. NOTAM FILE CXO.
RADIO AIDS TO NAVIGATION: NOTAM FILE CXO.
   INDUSTRY (L) VORTAC 110.2   IDU   Chan 39   N29°57.36' W96°33.73'   165° 19.0 NM to fld. 419/8E.

COMANCHE CO–CITY   (7F9)   2 NE   UTC – 6(– 5DT)   N31°55.01' W98°36.02'   SAN ANTONIO
1388   B   FUEL 100LL   L–15D
RWY 17-35: H3600X60 (ASPH)   S–17   LIRL
   RWY 35: Tree. Rgt tfc.
RWY 13-31: 2250X100 (TURF)
   RWY 13: Trees.
AIRPORT REMARKS: Attended 1400-2300Z‡ For fuel after hours call 915–356–3052.
COMMUNICATIONS: CTAF/UNICOM 122.8
   FORT WORTH FSS (FTW) TF 1–800–WX–BRIEF. NOTAM FILE FTW.
RADIO AIDS TO NAVIGATION: NOTAM FILE SJT.
   LAMPASAS (H) VORTACW 112.5   LZZ   Chan 72   N31°11.13' W98°08.51'   324° 49.7 NM to fld. 1290/08E.

COMMERCE MUNI   (2F7)   3 N   UTC – 6(– 5DT)   N33°17.63' W95°53.78'   DALLAS–FT. WORTH
516   B   L–13C, A
RWY 18-36: H3203X50 (ASPH)   S–13   LIRL   IAP
   RWY 18: Road.
AIRPORT REMARKS: Unattended.
COMMUNICATIONS: CTAF 122.9
   FORT WORTH FSS (FTW) TF 1–800–WX–BRIEF. NOTAM FILE FTW.
Ⓡ FORT WORTH CENTER APP/DEP CON 127.6
RADIO AIDS TO NAVIGATION: NOTAM FILE FTW.
   SULPHUR SPRINGS (L) VOR/DME 109.0   SLR   Chan 27   N33°11.92' W95°32.56'   280° 18.7 NM to fld.
   480/8E.   VOR voice out of svc indefinitely.

CONIS   N32°46.48' W96°46.51'   NOTAM FILE DAL.   DALLAS–FT WORTH
NDB (LOM) 275   LV   311° 5.8 NM to Dallas Love Fld.

CONOR   N27°50.08' W97°34.60'   NOTAM FILE CRP.   BROWNSVILLE
NDB (HW/LOM) 382   CR   127° 5.6 NM to Corpus Christi Intl   H–5B, L–16F

Fig. 8-6. A page from the South Central U.S. Airport/Facility Directory showing various airport information, including geographical coordinates, runway information, and communications frequencies. (For illustrative use only—not for navigation use.)

NOAA
N/CG33-Distribution Branch
Riverdale, MD 20737
and most FBOs and many mail-order chart services.

LORAN/GPS Navigator Atlas—a *Sky Prints* publication updated every eight weeks
(Fig. 8-7). It is a complete ensemble of charts in a spiral binder designed specifically for the pilot using LORAN or GPS and includes a listing of geographical coordinates for VORs, NDBs, LOMs, airports with paved runways longer than 2500
feet, and approach and low altitude fixes. The *LORAN/GPS Navigator Atlas* is
available from:

Air Chart Systems
13368 Beach Ave.
Venice, CA 90292
(800) 338-7220

Fig. 8-7. Sample of *Sky Prints*
LORAN/GPS Navigator Atlas
showing airport names,
identifiers, and geographical
coordinates. Note that starting in
April 1994, geographical
coordinates will be listed to the
hundredths of a second. (For
illustrative use only—not for
navigation use.) Air Chart Systems

WYOMING
AFTON Muni (AFO) 042-43.2N/110-56.0W
ALPINE (46U) 043-11.1N/111-02.5W
BIG PINEY-Marbleton (BPI) 042-35.1N/110-06.5W
BUFFALO Johnson Co (BYG) 044-22.9N/106-43.3W
CASPER Natrona Co Intl (CPR) 042-54.5N/106-
    27.8W
CHEYENNE (CYS) 041-09.3N/104-48.8W
CODY Yellowstone Reg (COD) 044-31.2N/109-01.4W
COKEVILLE Muni (U06) 042-02.7N/110-58.0W
COWLEY/LOVELL/BYRON North Big Horn Co (U68)
    044-54.7N/108-26.6W
DIXON (9U4) 041-02.3N/107-29.8W
DOUGLAS Converse Co (DGW) 042-47.8N/105-
    23.2W
DUBOIS Muni (U25) 043-32.9N/109-41.4W
EVANSTON-Uinta Co Burns Fld (EVW) 041-16.5N/
    111-01.9W
FT BRIDGER (FBR) 041-23.6N/110-24.4W
GILLETTE-Campbell Co (GCC) 044-20.9N/105-
    32.4W
GREYBULL South Big Horn Co (GEY) 044-31.0N/
    108-05.0W
GUERNSEY Camp Guernsey (7V6) 042-15.6N/104-
    43.7W
JACKSON Hole (JAC) 043-36.4N/110-44.3W
KEMMERER Muni (EMM) 041-49.5N/110-33.5W
LANDER Hunt Fld (LND) 042-48.9N/108-43.8W
LARAMIE General Brees Fld (LAR) 041-18.8N/105-
    40.5W
LUSK Muni (LSK) 042-45.0N/104-23.9W
NEWCASTLE Mondell Fld (ECS) 043-53.1N/104-
    19.1W
PINE BLUFFS Muni (82V) 041-09.2N/104-07.8W
PINEDALE Ralph Wenz Fld (PNA) 042-47.7N/109-
    48.4W
POWELL Muni (POY) 044-52.1N/108-47.6W
RAWLINS Muni (RWL) 041-48.3N/107-12.0W
RIVERTON Reg (RIW) 043-03.8N/108-27.6W
ROCK SPRINGS-Sweetwater Co (RKS) 041-35.6N/
    109-03.9W
SARATOGA Shively Fld (SAA) 041-26.7N/106-49.5W
SHERIDAN Co (SHR) 044-46.4N/106-58.6W
THERMOPOLIS Hot Springs Co-Thermopolis Muni
    (THP) 043-39.5N/108-12.8W
TORRINGTON Muni (TOR) 042-03.9N/104-09.2W
WHEATLAND Phifer (EAN) 042-03.3N/104-56.0W
WORLAND Muni (WRL) 043-57.9N/107-57.0W

Aviator's Data Log—a complete listing of all U.S. airports, public and private, with airport information, geographical coordinates, ID decoder, VORs, NDBs, and an alphabetical cross reference by location. The *Aviator's Data Log* is available from:

AeroNautical Products
P.O. Box 90349
San Diego, CA 92169
(800) 801-6187

JeppGuide Airport Directory—a Jeppesen publication up-dated every 120 days. Containing complete airport information and runway diagrams, it is available by geographical area of the country—Northeast, Southeast, South Central, Great Lakes, and Western. The *JeppGuide Airport Directory* is available from:

Jeppesen Sanderson
55 Inverness Drive E.
Englewood, CO 80112
(303) 799-9090

AOPA's Aviation USA—an annual publication from the Aircraft Owners and Pilots Association. It includes geographical and other airport information, along with many airport runway diagrams. The *AOPA's Aviation USA* is available from:

AOPA
421 Aviation Way
Frederick, MD 21701
(301) 695-2000

Many types of charts, and other flying paraphernalia are available from:

Sporty's Pilot Shop
Clermont County Airport
Batavia, OH 45103
(800) 543-8633

Portable aviation GPS receivers

The GPS portable meant for aviation use will have a permanent database included in the package. The database will contain, at a minimum, the geographical coordinates for all public airports with runways paved and more than 2500 feet long (Fig. 8-8).

A few portables have more extensive databases that include all airports, communication frequencies, and more. Some portable GPS receivers have E6B func-

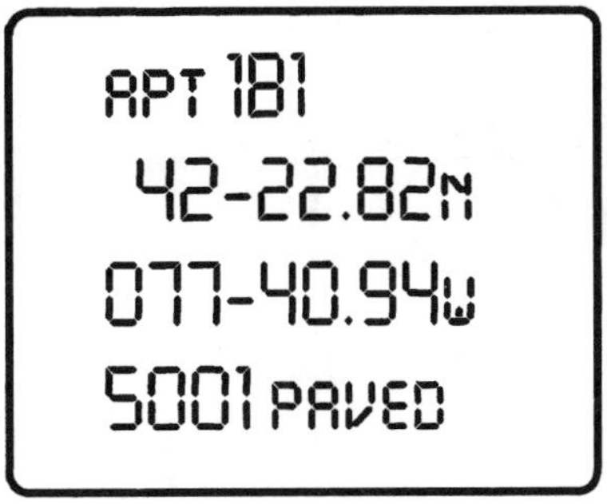

Fig. 8-8. Example of a simple database inquiry showing the airport's identifier (1B1), geographical coordinates, and paved runway length. (For illustrative use only—not for navigation use.)

tions for calculating winds, speeds, times, etc. They are quick and efficient, requiring but a few keystrokes for data entry.

One database function which is particularly important to the pilot is the *nearest airport* feature. At the press of a single button, or simple combination of buttons, the receiver determines and displays the nearest airports, usually ten nearest, to your current position. This is great in an emergency and fun when looking for new airports at which to land.

Moving-map portables

In chapter 9 you will learn about moving maps used with upper-level GPS receivers and add-ons. Although sounding very complicated and high priced, many of the features found on very expensive GPS receivers and computer-based add-ons can be found on some portable GPS receivers. To ascertain what receivers display moving maps, check in chapter 10 under "Hand-held GPS receivers."

Moving maps on portables generally display all of the normal alpha/numeric information and include a graphic display of airport information with a runway diagram, an improved CDI, and location of nearby facilities (Figs. 8-9 and 8-10).

Portable GPS receivers with moving maps are, at times, more difficult to read than their expensive panel-mounted brothers, however, they offer an affordable entry into the hi-tech world of visual satellite navigation.

Caveats for portable GPS receivers

There are certain limitations for using a portable GPS receiver in an airplane. If these limitations are not recognized and compensated for, the navigation results will be dismal. These apply to all portable GPS receivers.

First and foremost, the portable GPS should be equipped with a remote antenna. The receiver will not work well inside the airplane as the satellite signals cannot penetrate the plane's hull. Typically, remote antennas are supplied with hand-helds along with a suction cup style antenna mount for attachment to the inside of the windshield. Some users go so far as to have a permanently-mounted GPS antenna installed on the top of their plane.

A sharp maneuver of the aircraft may cause momentary shadowing of the signal and loss of GPS information. This can vary from a change of 3D to 2D posi-

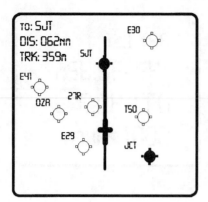

Fig. 8-9. This moving-map graphical display shows a plane on a course of 359 degrees 62 nm to San Angelo (SJT). Other facilities are indicated on the map including airports and VORs. (For illustrative use only—not for navigation use.)

Fig. 8-10. Example of a portable GPS moving-map receiver's database display showing the airport identifier, common airport name, runway information (two runways), lights, and communication frequency. To the right is an airport diagram. (For illustrative use only—not for navigation use.)

tioning down to a complete loss of data, depending upon the satellites in use, their location in relation to the airplane, and how they can become shadowed by the surfaces of the airplane (Fig. 8-11).

The electric power used by portable GPS receivers is normally in the form of an AA battery pack. As a cost-saving measure, rechargeable NiCad batteries make a good investment. They can be charged hundreds of times. However, NiCad batteries exhibit a shorter use period than non-rechargeable batteries, typically about half the stated battery life.

Direct connection with the airplane's electrical system eliminates the continuing expense of battery purchases. A feature to look for on a portable receiver is a means of automatically switching to internal-battery power when external power from the airplane's electrical system is lost; a nice safety feature.

Full-battery operation GPS gives space-age navigational capabilities to planes with no electric system; including some classics and many homebuilts. However, let me again point out that batteries are expensive and do not last very long.

Recognize that the portable GPS receiver is just that—portable. It is not fastened to anything on a permanent basis, and can therefore move about. In an airplane such movement can cause pilot confusion from situations such as bending to

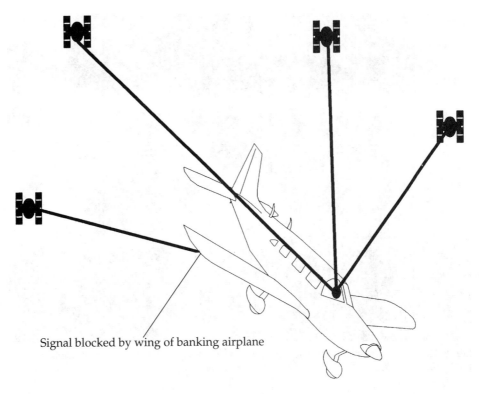

Signal blocked by wing of banking airplane

Fig. 8-11. During maneuvers, the satellite signals can be shadowed (blocked) by portions of the airplane.

retrieve the instrument from the floor, not knowing the location after it perhaps slides under the seat, or damage to the unit if, as an example it falls and may be crushed under a rudder pedal. Flying with a loose piece of equipment moving around the cockpit is not recommended—it is not efficient, it is inconvenient, and it is downright dangerous.

To alleviate the problems of using a portable GPS receiver in a plane, some manufacturers have designed panel or yoke mounts for their products. Trimble's *Flightmate* and Garmin's *55* and *95* portables, for example, have mounts designed to hold the receiver to the control yoke (Fig. 8-12). Mounts that fasten to the panel to hold the receiver are also available. After all, not all airplanes use control yokes.

One manufacturer, Motorola, uses a lock-in mount setup which makes power and external antenna connections for you (Fig. 8-13). This is very neat, from an aesthetics point of view, and very simple for the user-mounting and hookup system.

For GPS hand-helds totally lacking an adequate mount, AeroNautical Products produces the *Universal GPS Receiver Clamp*. This device is adaptable to a number of portables and clamps onto the instrument panel or side walls (Fig. 8-14).

No matter what make or model portable GPS receiver you have, remember to remove it from the plane when it is not in use. They are small, popular devices, and very prone to theft.

Fig. 8-12. This Flightmate is mounted to the control yoke, providing excellent visibility to the pilot.

PANEL-MOUNTED GPS RECEIVERS

Panel-mounted GPS receivers function the same as the portables. The only difference is they are mounted and permanently connected to the airplane's electrical system and to a mounted external antenna. There are no unsightly wires or finicky mounts to deal with. The only problem can be in locating the unit physically on the panel, particularly if the stack is already full.

It is not advisable to remove any functioning conventional navaid equipment to make room for GPS. It will not be cost effective, as used VHF navigation equipment—or any used aviation radio equipment in general—has little cash value. As a side note, when purchasing used equipment the reverse is not true—that same used equipment will suddenly gain in value. Further, recognize that GPS has not been fully approved for use under all flight circumstances.

Display placement

Due to the amount of information the GPS unit is capable of displaying, the receiver should be positioned for easy reading by the pilot. None have the display on the yoke, as do some portables. None are easily removable for use in another airplane. The best position for pilot visibility is at the top of the radio stack, which often may be inconvenient and expensive to accomplish.

All panel-mounted aviation GPS receivers currently on the market have built-in aviation databases offering varying features.

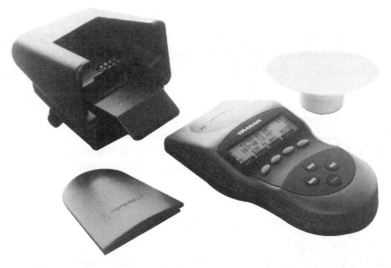

Fig. 8-13. Motorola's TRAXAR GPS receiver with the lock-in mount and external antenna. Motorola

Fig. 8-14. The Universal GPS Receiver Clamp allows convenient placement of portable GPS receivers.

Databases

Generally speaking, the databases used on panel-mounted GPS receivers are more extensive than those found on portable receivers. It is the norm to find databases providing the coordinates for all airports, NAV-aids, fixes and specific data about individual airports including: communications' frequencies, hours of operation, FBO, fuel, etc. (Fig. 8-15). Note, however, that some databases are so extensive that they can be confusing to the user, as they take too much time to use and read. Readability is a function of the visual presentation made of the information, a design factor of the particular receiver. Remember, GPS is for navigation and that other sources can be checked for information of lessor importance (i.e., what brand of fuel is sold at an airport, etc.).

NAVIGATING WITH GPS

Learning how to navigate with GPS is really quite simple. The receivers do all the work, and the operator needs only to learn how to use a particular make/model. For purposes of writing this chapter several different panel and hand-held receivers were used. In general, most GPS receivers operate very similarly to one another, providing the same high-quality information necessary for accurate navigation.

The largest difference between the various receivers used was the means of inputting information into the receiver's processor. Some operate via a keyboard resembling that found on a telephone while others use point-and-shoot on screen menus. I was comfortable with either and really can recommend no preference (Fig. 8-16). Note, that a backlighted keyboard is excellent for use at night.

Customizing your receiver

Each GPS receiver can be customized for operation to suit the user. Normally this means changing the output units—feet to meters, MPH to kts, statute miles to nautical miles, etc.—to suit individual needs or desires.

One place to another

No doubt the greatest features of the GPS receivers is the ease of operation and the large amount of information output from them. All this is available with almost no effort from the user.

For example, you are flying a few miles west of Albany, New York, and decide to continue on to Rutland, Vermont. Just enter Rutland's identifier of RUT and you will be informed that Rutland is 69.3 nm on a bearing of 55 degrees (magnetic) (Fig. 8-17).

Note: Some receivers allow the common name of the airport to be entered, thus not requiring the pilot to memorize a number of identifiers. For example: Albany County, Albany, New York, identifier of ALB is easy to remember, however, Lycoming County, Williamsport, Pennsylvania, identifier IPT is not. Some are worse: 1H6, 95F, etc.

```
ALB Albany County Albany NY
ILS VOR VOR/DME ASR AV/JET
ATIS 120.45 TWR 119.50
GRND 121.70 APP 125.00
```

```
ALB Albany County Albany NY
ILS VOR VOR/DME ASR
010/190 7200 FT 100/280 5999 FT
AV/JET FBO
```

Fig. 8-15. These database samples display identifiers, common airport names and locations, approach information, runway numbers & lengths, communication frequencies, and fuel availability. (For illustrative use only—not for navigation use.)

Creating a waypoint

There are two basic means of creating a waypoint, assuming it is not included in the database being used. If you are going to use your current location as a waypoint you merely assign a name for the location and store it. If the waypoint you wish to enter is at some other location, enter the geographical coordinates for that location, give it a name, and store it. Either way, you will have created a USER waypoint. Other types of waypoints generally indicate what they are named for—VOR, NDB, APT, etc.—and are internal to the database.

The ability to enter the current location as a waypoint is particularly convenient when marking a location from the air. For example a geographical location not normally associated with flying (Fig. 8-18).

Using waypoints

The *navigate to* feature, found on all GPS receivers under various names—destination, GOTO, TO, etc.—is designed to take you from *here*—your present position, to *there*—whatever position or identifier you enter.

For routine operations involving navigation with GPS, this is the feature you will use most—and it is the easiest feature to understand.

It is via waypoints, TO and FROM, that we navigate from *here* to *there*—from the simple single-leg trip made by the selection of a single TO waypoint and computed from the present location to planned routes containing many waypoints.

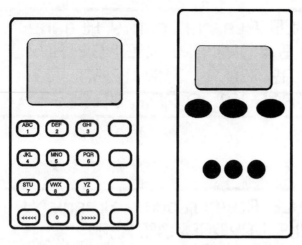

The keyboard on the left selects specific letters and numbers by pressing the key for that letter (8 for X) and pressing the arrow to move through V, W, X, and 8.

When the letter you desire shows, press ENTER. The keyboard on the right displays the letters on the screen and the key (s) indicated by the screen as arrow keys are toggled until the desired letters or numbers show. Then the key indicated as ENTER is pushed.

Fig. 8-16. Although these two GPS receivers have different user interfaces (keypads), they both operate in a similar manner. Neither simplicity nor complexity is an indication of capability.

Most GPS receivers allow for a simple one-button selection of the destination function, followed by entry of the destination name, generally entered by identifier (Figs. 8-19 and 8-20).

Routes are a group of waypoints entered into the receiver in the order you wish to fly them. Generally, there is little need for routes, except to fly around something. That something might be a particular area of airspace you wish to avoid, a mountainous area, or other location you wish to circumnavigate. The route may be envisioned as a series of dog-legs.

Following a coarse to the selected waypoint is monitored by statement and by CDI (Figs. 8-21 and 8-22).

Proximity alarms

Most GPS receivers will notify you when you are getting close to the desired waypoint, or in some cases, in the vicinity of a piece of airspace in which you do not belong. Unfortunately, unless you happen to be looking at the display when it is flashed, alarms may be missed. Many are in the form of audio beeping, which may not be heard in a noisy cockpit. Alarm data will remain indicated on the screen until acknowledged. Settings for range (distance) alarms are selected by the user—i.e., range 1 nm, range 10 nm, etc.

TO RUT
BRG 055 m
RNG 69.3 NM
ETE 00:25

Fig. 8-17. Present position and altitude examples. Notice the lower example indicates 2,306-foot altitude and 3D in the lower right corner. Altitude reporting is available only when the receiver is in the 3D (three-dimensional) mode. (For illustrative use only—not for navigation use.)

TO RUT 2306 FT
BRG 055 m
RNG 69.3 NM
ETE 00:25 3D

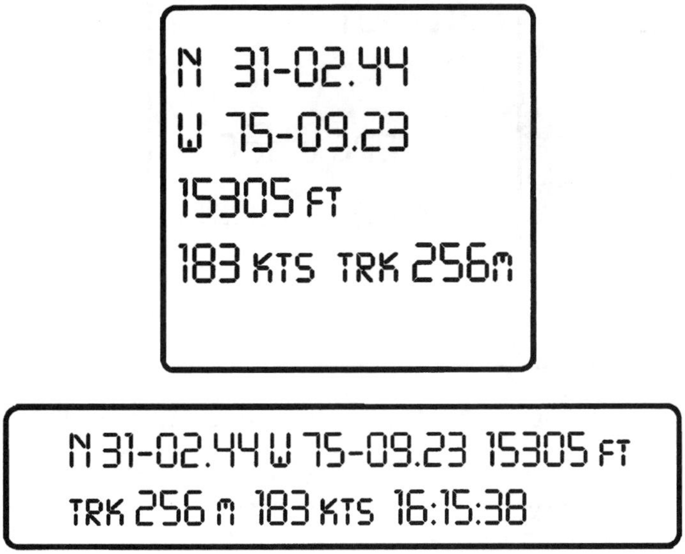

N 31-02.44
W 75-09.23
15305 FT
183 KTS TRK 256m

N 31-02.44 W 75-09.23 15305 FT
TRK 256 m 183 KTS 16:15:38

Fig. 8-18. At the press of a button these sample locations can be entered into the GPS receiver database as user waypoints. (For illustrative use only—not for navigation use.)

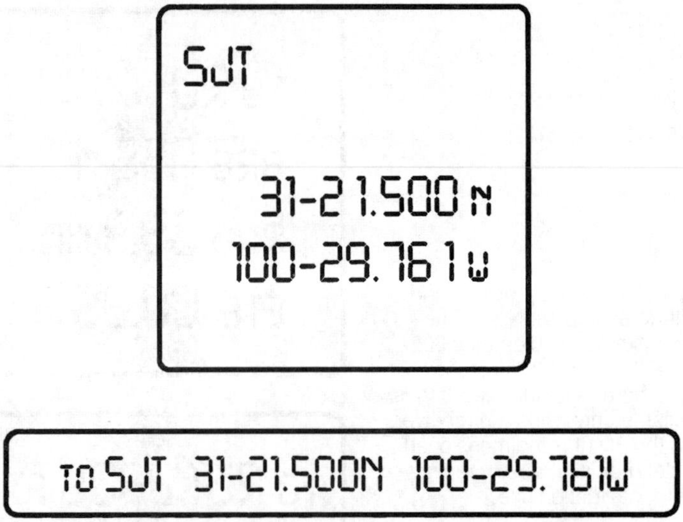

```
SJT



      31-21.500 N
   100-29.761 W
```

```
TO SJT  31-21.500N  100-29.761W
```

Fig. 8-19. Simple examples showing the waypoint SJT entered with the resulting geographical coordinates displayed in answer. (For illustrative use only—not for navigation use.)

```
GOTO SJT



    BRG 242m
    DIS 912 NM
```

```
TO SJT  31-21.500N  100-29.761W
BRG 242 m DIS  912NM ALT   2D
```

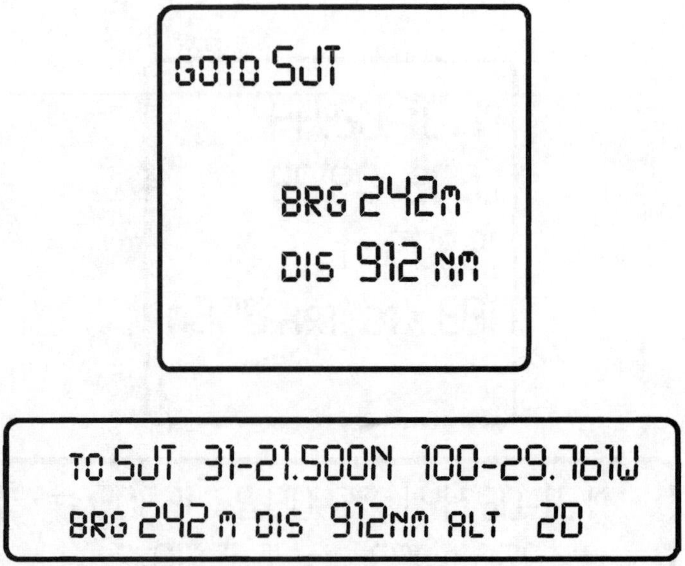

Fig. 8-20. Using the same waypoint (SJT) as the previous illustration, the bearing and distance are displayed in answer to the entry. Note the lower example indicates 2D for altitude—meaning only three satellites are being used by the receiver and the altitude cannot be calculated. (For illustrative use only—not for navigation use.)

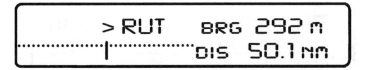

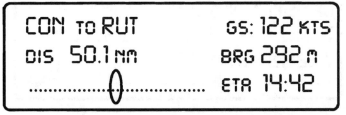

Fig. 8-21. Three examples showing various forms of enroute display, for the same time, location, and destination. (For illustrative use only—not for navigation use.)

Nearest waypoint

Most receivers built for aviation use have a feature called the *nearest facilities*. By using this feature, the display will indicate the bearing and distance to the nearest airport, VOR, NDB, FSS, etc. In an emergency, a GPS receiver can direct you to the nearest nine or ten (sometimes more) airports, a real time saver from having to pull out a chart and look at it (Fig. 8-23).

Other displays

Most GPS receivers are loaded with informational display capabilities. For example, nearly perfect time is available (Fig. 8-24).

Satellite reception information is shown depicting how many and what satellites are being tracked, their locations in the sky, and relative signal quality or strength (Fig. 8-25).

A final recommendation

Although the GPS is wonderful to use, highly accurate, and quite dependable—do not rely upon GPS alone! Use your senses and be aware of your surroundings. You can IFR (I follow the river), IFR (I follow the railroad), or IFR (I follow the road). In other words, keep your head out of the cockpit. Know where you are at all times.

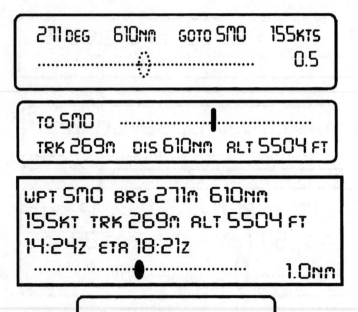

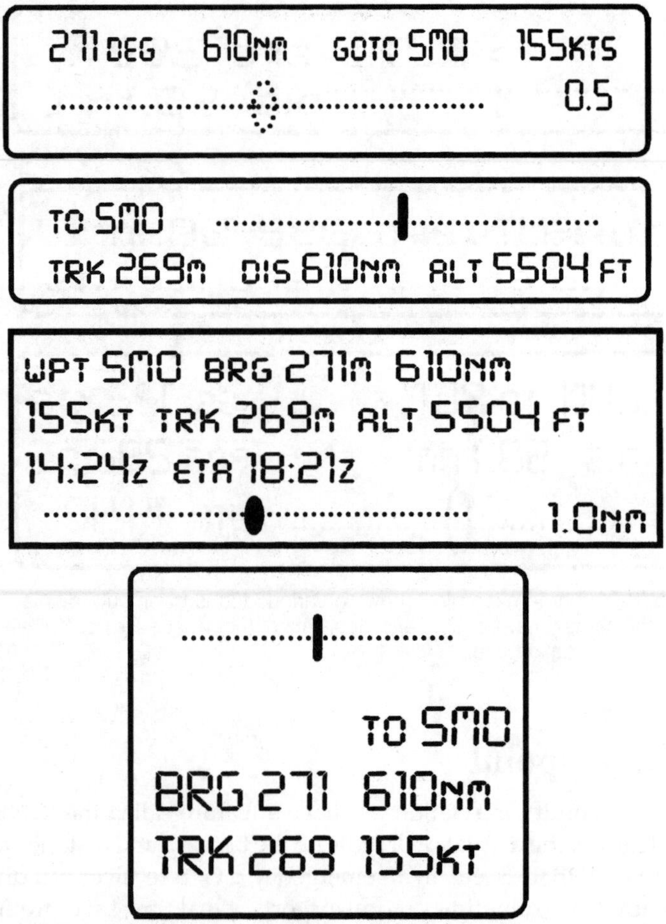

271 DEG   610 NM   GOTO SMO   155 KTS
······················⊕······················   0.5

TO SMO   ··················❙··················
TRK 269 m   DIS 610 NM   ALT 5504 FT

WPT SMO BRG 271 m 610 NM
155 KT TRK 269 m ALT 5504 FT
14:24 z ETA 18:21 z
··················●··················   1.0 NM

··················❙··················

TO SMO
BRG 271 610 NM
TRK 269 155 KT

Fig. 8-22. Four examples depicting the same location, time, and destination in different display formats. Top two examples are two line with CDI, third example is four line with CDI, and the bottom example is three line with CDI at the top. (For illustrative use only—not for navigation use.)

SJT NEAREST

BRG 271 m

RNG 11.2 NM

Fig. 8-23. Display showing the nearest airport (SJT, bearing 271 degrees, range 11.2 nm). (For illustrative use only—not for navigation use.)

Fig. 8-24. Display of the date
and time (as accurate as a $10
billion clock can be). (For
illustrative use only—not for
navigation use.)

```
LOCAL TIME
07-16-93


21:23:47
```

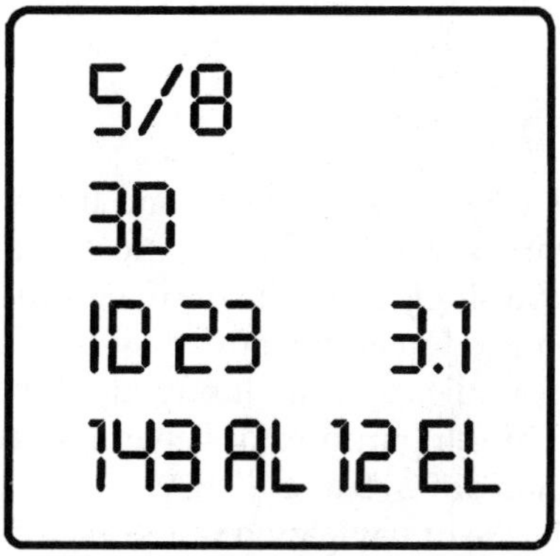

Fig. 8-25. Display showing the
number of satellites being
tracked (five) and the number in
view (eight), 3D positioning, the
PRN code of the specific satellite
(23), the signal quality (3.1), and
its location in the sky (azimuth
and elevation). (For illustrative
use only—not for navigation
use.)

```
5/8
3D
ID 23    3.1
143 AL 12 EL
```

9

Advanced airplane navigation with GPS

THE USE OF GPS is expansive and the system can be used as a part of an overall more complex integrated navigation system. Integrated systems use inputs from more than a single source to derive positioning information.

When used in an integrated system, a GPS receiver is referred to as a *sensor*. The same applies to other inputs used in the overall system, they are also called sensors. For the purposes of general aviation, the most common sensors used in combination are GPS and LORAN.

Multi-sensor navigation units, via a system of internal assessment, determine which sensor is providing the most suitable information for final display. Some output is based upon a combination of sensor inputs, which can include non-navigation information such as that provided by a fuel sensor, which provides display of fuel flow, fuel burn, supply on board, and calculated range and efficiency, altitude information taken from barometric source or Mode-C, or an air-data computer, which gives information displays of outside air temperature, true and indicated airspeeds, and winds aloft—manual data can be entered for calculation. Outputs from an integrated unit include CDI, HSI, moving map, and auto pilot.

The aviation words used to define a system of multi-sensor inputs and multiple outputs are the *flight management* or *navigation management* system.

LORAN and GPS

The popular combination of LORAN and GPS sensors into a single unit is the most available multi-sensor system (Figs. 9-1 and 9-2). These combination units are no larger in size than stand-alone GPS receivers and fit easily into a standard radio stack. The means of keeping the physical size down is remote installation of the sensors (generally the GPS component). Models in varying operational conformations include:

Fig. 9-1. Combination LORAN and GPS receiver in the form of an ARNAV FMS 5000 for general aviation.

Fig. 9-2. The ARNAV FMS 7000 flight management system for the upper end of general aviation and commercial applications.

ARNAV Systems

- ARNAV FMS 5000, 7000, and 9000 Multi-Sensor FMS
- II Morrow Apollo NMS
- Narco STAR-NAV 9000 (includes sensors for VOR and Glideslope)
- Northstar M2
- Trimble TNL-3000

Although general aviation primarily will be concerned with GPS/LORAN-based multi-sensor devices, other alternatives are available. These differing units utilize Omega in place of LORAN, due to the former's worldwide application (Fig. 9-3).

The common denominator among all of the multi-sensor units is the ease of operation, which is a signpost of nearly all GPS-based navigation. All use extensive databases, usually worldwide in scope.

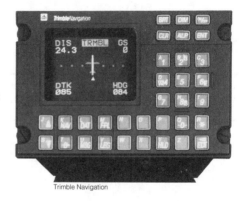

Fig. 9-3. Trimble TNL-7880 Omega/GPS-based navigation system.

Trimble Navigation

Note the TSOs

Multi-sensor and flight-management systems can receive TSO approval for parts of the overall system. The consumer is well advised to beware of advertising literature indicating approval under various TSOs that may or may not be appropriate for a particular use.

In the case of GPS, the appropriate TSO is C129 with its integrity and database requirements. Note that for IFR multi-sensor systems, the minimum performance standards are specified in TSO-C115a. Other pertinent TSOs include:

- TSO C44a for fuel sensors
- TSO C60b IFR LORAN use
- TSO C106 for air data computers

At no time is approval under one TSO an indicator of approval, or even pending approval, under another. For further discussion on TSO requirements specific to GPS refer to chapter 6.

MOVING MAPS

The most sought after display of GPS output is the electronic moving map. Moving maps apply the GPS PVT (position, velocity, and time) information to an extensive database and display the information on an electronic map.

Most GPS receivers used in aviation display navigation information in alphanumerics on line-oriented displays. The single deviation being the digital CDI that is common to most GPS receivers (Fig. 9-4).

The advantage of moving-map displays over any other method of presenting navigation data to the user is the direct application of the *you are here principle.* The map displays typically show the airplane's location in reference to facilities—VOR, airports, etc.—and airspace boundaries—Class B and C, formerly called TCA and ARSA—complete runway diagrams, and alphanumeric display of navigation and database information. Generally, the scale, or area of coverage, can be changed to suit the immediate needs of the pilot.

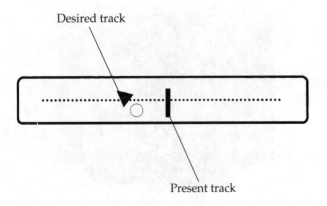

Desired track

Present track

Fig. 9-4. Example of the familiar electronic CDI found on most alphanumeric-display GPS receivers. Steer towards the desired track to correct cross track error.

Types of moving-map systems

Moving-map displays come in various configurations from one-piece units to third-party (manufactured by other than the receiver's manufacturer) moving-map displays. The latter are the more common means of supplying the pilot with map-type information.

Complete receiver/display panel-mounted units such as the AlliedSignal KLN 90 GPS receiver, contain an internal moving-map display as part of the over-all GPS receiver unit. Specifications and a photo of the KLN 90 may be seen in chapter 10.

Portable GPS receivers are also available with moving-map displays and cost considerably less—in the $1500 range. Excellent examples are the Garmin *GPS 95* and the Sony *IPS-760-AVD*, both, featured in chapter 10.

Third-party moving maps depend upon connection with a GPS receiver for PVT data, however, the receiver may be in the form of a sensor rather than a full-featured receiver with a complete display and database. Moving-map displays use their own databases and do not depend upon the GPS receiver's database. As with all databases, they, too, require up-dating—at a cost. Third-party maps come in many forms:

- Panel-mounted remote displays
- Hand-held display
- Laptop computer
- Chart based display

There are advantages and disadvantages to each. For example, panel-mounted moving-map displays offer solid and permanent positioning, remote from the receiver. The display, large instrument size, can be installed nearly anywhere the pilot desires, affording excellent viewing (Figs. 9-5 and 9-6). The disadvantage is finding the panel space for the installation.

Obvious advantages of a portable system are no lost panel space, no complicated (read expensive) installation, light weight, and the ability to operate on bat-

Fig. 9-5. The ARGUS 3000 moving-map display for general aviation can interface with nearly any GPS receiver.

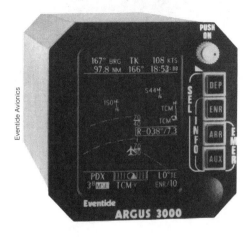

Eventide Avionics

Fig. 9-6. The ARGUS 5000 moving-map display for commercial aviation can interface with nearly any GPS receiver.

Eventide Avionics

tery power when desired. In general, the portable units are quite capable and offer a wide array of features (Fig. 9-7). A disadvantage of some portable units is a small display screen.

Laptop computer-based moving-map displays offer the same advantages as portable displays and have extensive computing capabilities in addition. The computer-based system can be attractive as many pilots already have a GPS receiver in their airplane and own a computer. A connection between both, and the installation of proper software, is all that is required to get the advantage of a moving map. The display will generally be quite large (computer screen limited only) (Figs. 9-8 and 9-9).

Fig. 9-7. MEMTEC co-pilot portable moving-map display. This lightweight device can be held to your leg with a piece of Velcro.

Memtec Corp.

Fig. 9-8. The Peacock LapMap moving-map software system for laptop or palmtop computers.

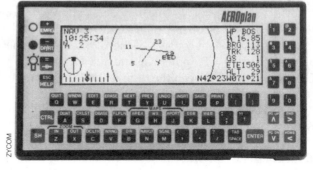

Fig. 9-9. The ZYCOM AERO plan unit offers many hard key features and an excellent display.

On the down side of the laptop computer-based system, you will have a relatively large unattached item floating around the cabin with wires connected to other items in the same cabin. An unsafe condition will occur if this equipment is allowed to distract a pilot from the duties of flying. This problem may be somewhat alleviated in the future by a removable LCD computer display that could be placed in a secure (fastened down) location.

The chart-based moving-map display uses an electronic overlay and standard air-navigation charts—WAC, sectionals, etc. The immediate advantage of this system is the use of charts the pilot is already familiar with, and not just electronic images drawn on a screen (Fig. 9-10).

Fig. 9-10. The TELDIX CoPilot moving map superimposes positioning information onto standard aviation charts.

Laptop computer navigation

Laptop computers have worked their way into everyday life for many of us, so it stands to reason there would be a place for them in GPS aviation applications.

For GPS-based navigation, a software package allows a receiver to communicate with a small computer. The software package interprets the GPS PVT information and applies it in the manner the programmer intended. For our purposes, this will be visual output in the form of a moving map, alphanumeric information, or a combination of both.

The moving-map system may be as simple or complete as the designer desires. It may also be operator-limited to some planned extent.

The basic system consists of the GPS receiver, whether a complete receiver or only a sensor, a GPS antenna, and the laptop computer with proper software. Communication between the receiver and the computer is via a RS232 port (fancy computer talk for a plug).

Most laptop-computer/GPS map systems will make E6B calculations such as winds aloft, true air speed, wind components, etc.

An obscure use, however one that could prove to be very important, is the ability of some computer-based systems to record a flight's path. The noted times and geographical locations could later be used to challenge potential airspace violations.

Advanced forms of laptop-map systems are capable of reading and displaying information input from remote sensors that monitor the airplane's operation: electric system voltage, fuel usage, EGT, OAT, etc. (Fig. 9-11).

Using a moving map

The moving-map display shows all the information a standard alphanumeric display can, then visually displays locations along with specific alphanumeric information. The following is based upon information and illustrations courtesy of Eventide Avionics as provided in their Argus 3000 Reference Manual.

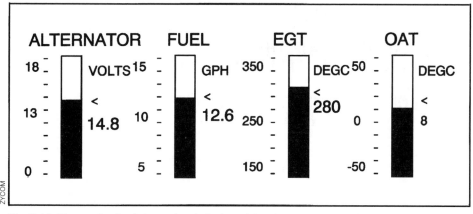

Fig. 9-11. Non-navigation informational display of the electrical system, fuel usage, exhaust gas temperature, and outside air temperature.

A Sample First Flight with the Argus 3000:

As an introduction to the Argus 3000, let's take a short flight. Seat belts and shoulder harnesses fastened?

When you turn on the Argus, the first thing you see after a short display test is the "LIMITATIONS" screen. Please read it now (Fig. 9-12).

After the Argus passes its self-test, a second screen (Fig. 9-13) giving information about the unit is displayed; it reminds you that you must have your charts with you to be legal. This screen also tells you the expiration date of the database. Once you have pressed any button, one of two messages that warn you of difficulties with the GPS may appear. If the GPS hasn't been turned ON, the **LRN DATA LOST** message will appear (Fig. 9-14). If the GPS is ON but hasn't yet "acquired," you will see the **LRN NAV INVALID** message (Fig. 9-15). Normally these messages will disappear quickly, but they are vital to alert you to system problems and you should be aware of their existence.

So much for the administrative stuff, let's go flying! Our trip today will be from N07 (Lincoln Park, NJ) to ABE (Allentown-Bethlehem, PA), a trip of some 54 nm.

Assuming that the GPS is working and has acquired, the Argus should show a display of the bearing and distance to the waypoint (ABE), but the graphic display area shows **NO HEADING OR TRACK** (Fig. 9-16 on page 144). That's because we're not moving, so there is no track or groundspeed. The mode shown is **PLAN**, because that's always the first display seen. This mode "autoranges" so that the most distant waypoint is always shown at the edge of the screen.

As we begin taxiing to runway 19 (ground TK 010°), the **PLAN** display appears, showing the destination to our left (Fig. 9-17). Depending on the GPS, this display may appear sooner or later during taxiing. It may also appear somewhat erratic while we're on the ground since our speed will be low and the GPS may have difficulty following our tight turns while taxiing. From the **PLAN** mode we can determine how to proceed. We'll be taking off on runway 19, so it would seem that a right turnout would be the best procedure. Note

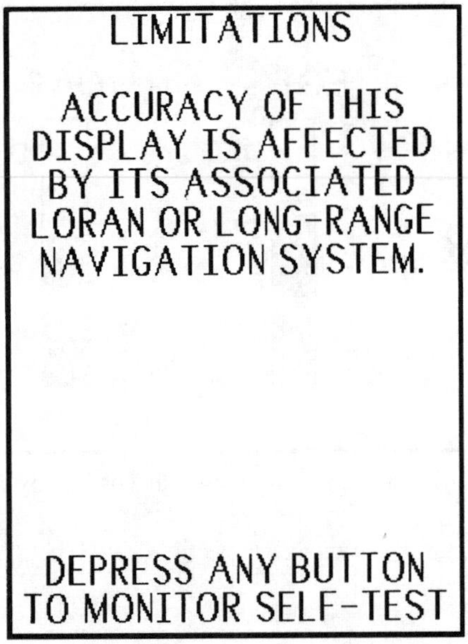
LIMITATIONS
ACCURACY OF THIS DISPLAY IS AFFECTED BY ITS ASSOCIATED LORAN OR LONG-RANGE NAVIGATION SYSTEM.
DEPRESS ANY BUTTON TO MONITOR SELF-TEST
Eventide Avionics

Fig. 9-12. Limitations screen. (For illustrative use only—not for navigation use.)

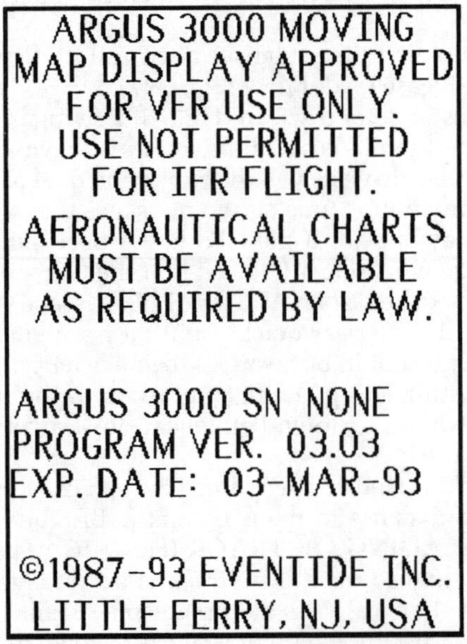
ARGUS 3000 MOVING MAP DISPLAY APPROVED FOR VFR USE ONLY. USE NOT PERMITTED FOR IFR FLIGHT.
AERONAUTICAL CHARTS MUST BE AVAILABLE AS REQUIRED BY LAW.
ARGUS 3000 SN NONE
PROGRAM VER. 03.03
EXP. DATE: 03-MAR-93
©1987-93 EVENTIDE INC.
LITTLE FERRY, NJ, USA
Eventide Avionics

Fig. 9-13. Self-test screen. (For illustrative use only—not for navigation use.)

that special use airspace is not shown in **PLAN** mode. When we go to **DEP**arture, we'll see that we're under the floor of the New York TCA and near two controlled airports. This will slightly affect the departure route.

As we switch to **DEP**arture, we're on the parallel taxiway heading for the threshold of runway 19 (Fig. 9-18) and the display depicts us heading for the threshold of runway 19. Note that at our current groundspeed (25 kts) we will

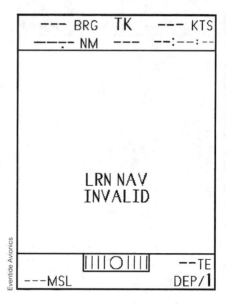

Fig. 9-14. GPS is off. (For illustrative use only—not for navigation use.)

Fig. 9-15. GPS has not been acquired. (For illustrative use only—not for navigation use.)

arrive at our destination in 2 hours and 10 minutes. Fortunately we won't be taxiing all the way. The CDI shows that we're on-course.

The next illustration (Fig. 9-19) shows us just becoming airborne from runway 19. Our speed has increased to 77 kts and our ETA is a more reasonable 42 minutes. We're still on-course.

Next, we see the airport behind us as we leave the local traffic area (Fig. 9-20). Note that we're slightly off-course because of the 45 degree turnout, and

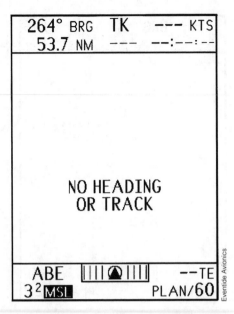

Fig. 9-16. No heading or track. (For illustrative use only—not for navigation use.)

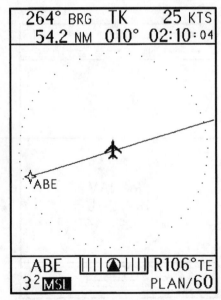

Fig. 9-17. Destination is at the left and the range scale is set to 60 nm. (For illustrative use only—not for navigation use.)

our speed is up to 90 kts. Also, note that our track error has decreased from 74 degrees to 24 degrees since we made our right turn.

We've left the airport and switched to the 5 mile **DEP**arture range (Fig. 9-21). To our left is CDW, a controlled airport we want to avoid. To our right is the New York TCA outer boundary. The legend shows us we must avoid altitudes between 7000 and 3000 feet. We are now heading on-course.

In the next example (Fig. 9-22), we have switched to the **ENR**oute mode.

Eventide Avionics

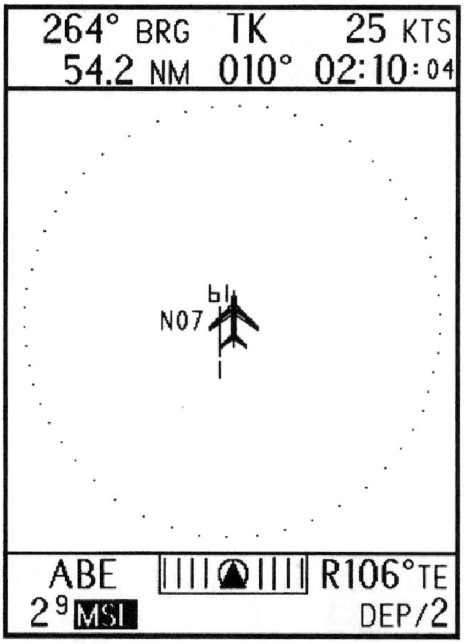

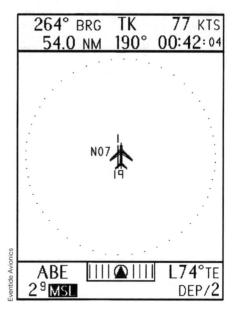

Fig. 9-18. Note the separation between the airplane and the runway. (For illustrative use only—not for navigation use.)

Eventide Avionics

Fig. 9-19. Takeoff from runway 19. (For illustrative use only—not for navigation use.)

Our position hasn't changed, we've just switched ranges. The 20-mile range shows more of the TCA and, of course, a greater distance ahead of the aircraft.

A bit closer to our destination (Fig. 9-23). Let's confirm the Argus display with our VOR/DME. (It is always a good idea to cross-check our Nav sources!) By pressing **SEL**ect, BWZ VOR is presented. According to the Argus, we are on radial 083° at a distance of 17.5 nm.

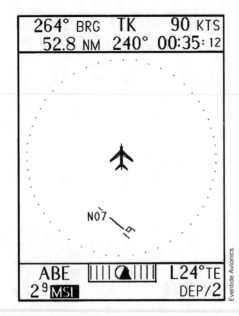

Fig. 9-20. Moving away from the airport after takeoff . (For illustrative use only—not for navigation use.)

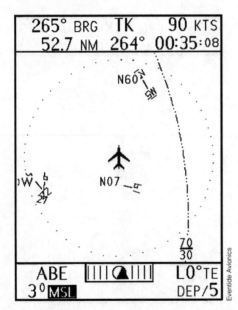

Fig. 9-21. Five-mile range shows nearby airspace to be avoided. (For illustrative use only—not for navigation use.)

Pressing **INFO** gives us the frequency of the Broadway (BWZ) VOR (Fig. 9-24). We can confirm VOR data by comparing the VOR/DME to the Argus. The numbers should be very close except, of course, at station passage, when the radial on both the Argus display and the VOR indicator will change rapidly. (Never use the Argus for VOR approaches!) ABE and the surrounding ARSA are now visible as we switch to **ENR** 30 (Fig. 9-25).

Eventide Avionics

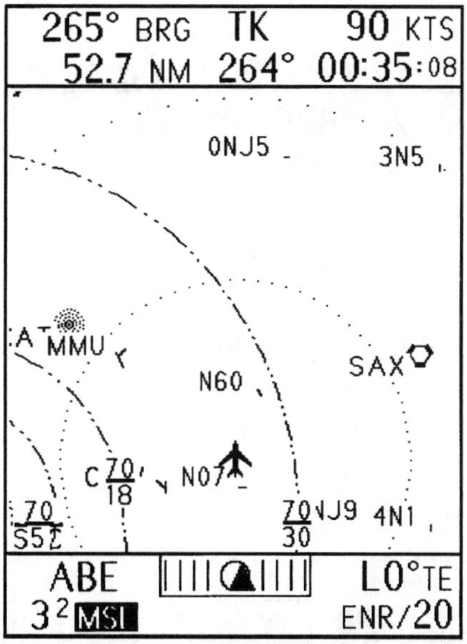

Fig. 9-22. Twenty-mile range showing additional surrounding area information (For illustrative use only—not for navigation use.)

Eventide Avionics

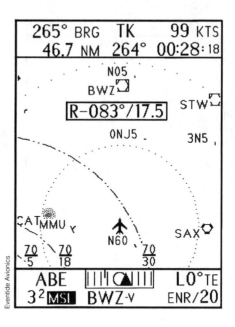

Fig. 9-23. Checking the bearing/range to VOR BWZ—displayed in the box. (For illustrative use only—not for navigation use.)

In another five minutes we're within hailing distance of ABE approach. We **SEL**ect the airport this time (**ABE-A**, not **ABE-V**) and note that the name of the facility appears briefly above the lower window (Fig. 9-26).

Press **INFO** to find the ATIS, tower, and approach frequencies (Fig. 9-27).

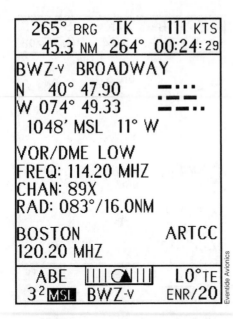

Fig. 9-24. BWZ VOR alphanumeric information display. (For illustrative use only—not for navigation use.)

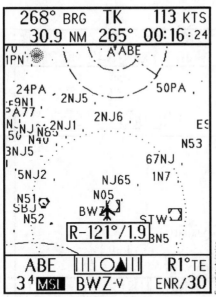

Fig. 9-25. Destination appears at the top of the display. (For illustrative use only—not for navigation use.)

Press **INFO** twice until the **PLAN VIEW** display appears. This display comes up after the last available runway information. Note that we are heading straight toward the airport and are almost aligned with runway 24 (Fig. 9-28).

After listening to the ATIS, we call approach and switch back to **ENR**oute on the 10-mile range. Notice that we have penetrated the ARSA (Fig. 9-29).

We now switch to the **ARR**ival mode. In this case we are using the **AUTO-**

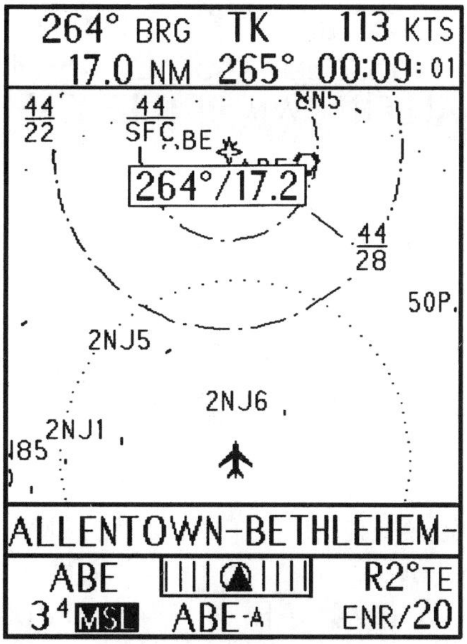

Fig. 9-26. Close to the destination. (For illustrative use only—not for navigation use.)

Eventide Avionics

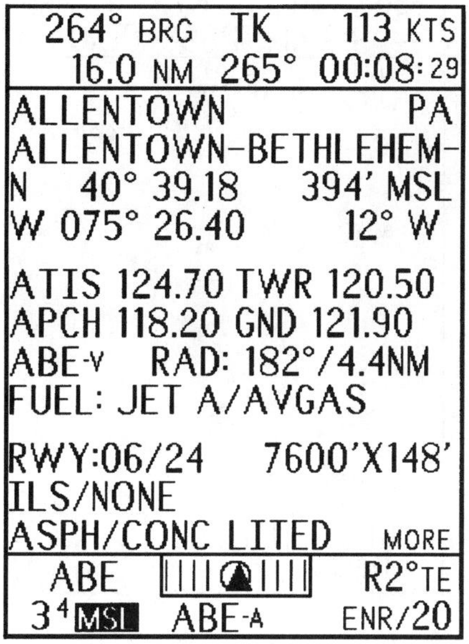

Fig. 9-27. ABE airport information including the communications frequencies. (For illustrative use only—not for navigation use.)

Eventide Avionics

ARRival feature, which allows the screen to automatically expand as we get closer to landing. (**AUTO-ARR**ival is distinguished by the display of **ARR** in reverse video.) Tower asks us to do a straight-in approach to 24, so we turn 40 degrees to the right to intercept the extended runway centerline (Fig. 9-30).

The screen has automatically switched to the 6 mile range, and we're al-

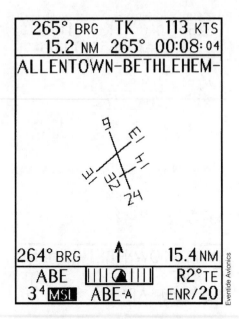

Fig. 9-28. Only 15 miles to go. (For illustrative use only-not for navigation use.)

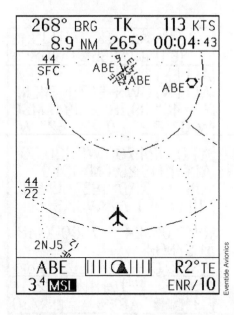

Fig. 9-29. Airspace penetration. (For illustrative use only—not for navigation use.)

most aligned with the runway. We're just crossing the 5 mile ARSA ring (Fig. 9-31).

Automatically set to 3 miles, we're on long final and nearly aligned with the runway (Fig. 9-32).

Finally, on the **ARR** 2 mile range, we're slowed down and cleared to land (Fig. 9-33).

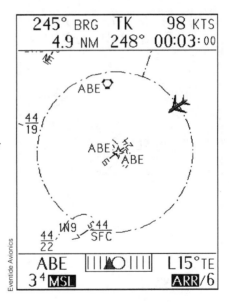

Fig. 9-30. Auto-expanding feature changes the range to eight nm. (For illustrative use only—not for navigation use.)

Eventide Avionics

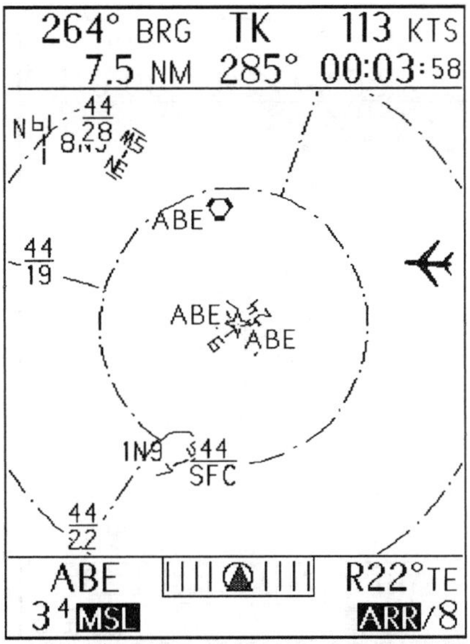

Fig. 9-31. Nearly lined up with RWY 24 at a six-mile range. (For illustrative use only—not for navigation use.)

Eventide Avionics

Moving maps allow the pilot to navigate around, therefore avoid, selected airspace (such as Class B & C). Prohibited, Alert, Warning, MOAs, ADIZ, and facility-associated airspace (Class B, C, & D) are generally included in the database and displayed on the screen (Fig. 9-34).

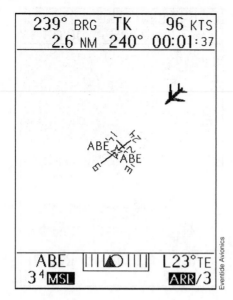

Fig. 9-32. On final for RWY 24. (For illustrative use only—not for navigation use.)

Eventide Avionics

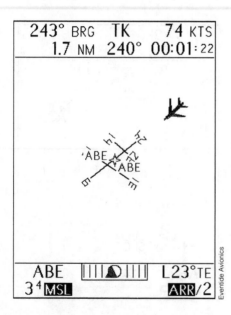

Fig. 9-33. Cleared to land. (For illustrative use only—not for navigation use.)

Eventide Avionics

HELP—EMERGENCY! In the emergency mode a display of airports with bearing/range information and radio frequencies (if applicable) is shown. Generally the search can be limited to a specified range (i.e., 25 nm) (Fig. 9-35).

Database notes

Moving-map displays are only as good as the information contained in the database. The more extensive the database, the more complete the map display.

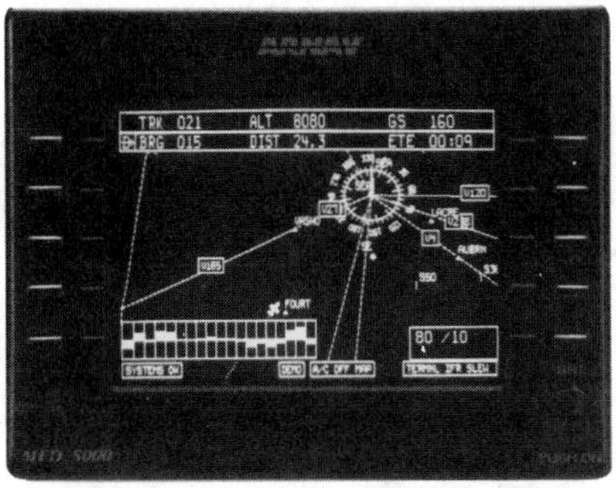

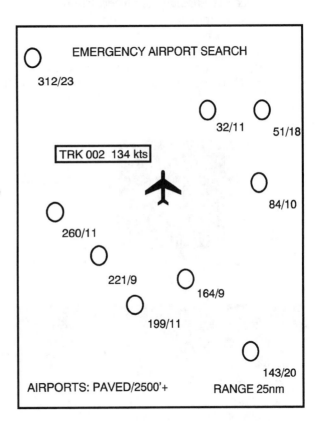

Fig. 9-34. Airway display on ARNAV's MFD 5000 4.25-×-3-inch CRT display.

ARNAV Systems

EMERGENCY AIRPORT SEARCH

312/23

32/11   51/18

TRK 002 134 kts

84/10

260/11

221/9   164/9

199/11

143/20

AIRPORTS: PAVED/2500'+   RANGE 25nm

Fig. 9-35. In an emergency, the moving map can display nearby airports with bearing and range information. Track and speed are in the box.

Settle for nothing less than all features including airspace alerting, which can prevent violations. Remember: the database must be up-dated periodically, depending upon use, at a cost.

Additionally, the map display must have a high ratio of selectable ranges. Typically 1 mile to 200 miles or more. These selectable ranges allow for close-up looks near destinations and wide area maps while enroute.

Receiver purchasing note

If you are planning to purchase a moving-map display, assure yourself that it will be compatible with your GPS receiver. Note that some displays offer a non-panel-mounted GPS sensor unit designed to operate with their specific display. These sensors are usually OEM—original equipment manufacturer—by a known producer. For example: ZYCOM uses the Rockwell International NavCor V GPS receiver.

CDIs

GPS receivers display their information output via digital displays in either alphanumeric or graphic form. None directly display anything similar to a traditional CDI (refer back to Fig. 9-4).

Most panel-mounted GPS receivers have provisions for sending information to outboard display devices, such as an analog CDI. For these purposes, a panel-mounted CDI is appropriate and is familiar from VOR use (Fig. 9-36).

Fig. 9-36. Panel-mounted CDI for use with GPS.

Fig. 9-37. Trimble TNL-2800 base-station ensemble showing the computer tracking of flight progress.

GPS LANDING SYSTEMS IN THE FUTURE

GPS currently provides the approved capabilities for nonprecision approaches (see chapter 6) and has the potential for further instrument approach use. As mentioned in earlier chapters, the FAA is currently investigating these possibilities.

As an experiment with precision GPS landing systems, Trimble Navigation installed a TNL-2800 Differential Global Positioning System Landing and Tracking System at Oshkosh in the summer of 1993 for demonstration purposes during the annual Experimental Aircraft Association convention (Fig. 9-37).

Using DGPS as the means to make GPS more accurate, the goal of the system was to meet Category I IFR requirements. In the aircraft, CDI, GSI, and HSI are used as displays. Moving-map displays are also usable with the system.

Unlike the DGPS signals used in other services, the aviation signal is VHF, therefore limited to line of sight. The accuracy should be in the one to three meter range.

A side feature of the system is a radarless radar. Through a bidirectional data link, the locations and progress of up to 24 aircraft can be monitored on a computer screen.

At this time the FAA continues to investigate GPS for Category I, II, and III approaches, with a final determination to be made in the future.

10

An overview of aviation GPS receivers

BEFORE MAKING A DECISION about the purchase of a GPS receiver for flying use, or any other use for that matter, there are a number of things to consider, over and above the well known abilities of the system.

You want to purchase not only a technically capable receiver, but you want a receiver that will remain usable for a number of years to come. Because GPS is still very new, you must examine your specific requirements and compare them to the equipment currently available—with an eye on what new equipment may be offered in the near future.

Databases

Minimally, a standard GPS receiver's aviation database contains the latitude/longitude coordinates of most public airports and navaids—VORs and NDBs—and is stored in permanent memory.

Some, but not all, simple-operating schemes require the user to know the three- or four-character designators for the facility of interest—SJT for San Angelo, Texas, for example—while more complete systems store database information not only by a designator, but by common facility or city names.

Although a preprogrammed database is standard for all aviation-type GPS receivers, they do have drawbacks—of which updating is the largest. Some GPS receivers use databases that are easily user updated, while others require the receiver to be returned to the factory for service. In general, hand-held units are not user friendly regarding updates. Most panel-mounted receivers have user-friendly database cards, easily installed/removed and exchanged for newer ones.

Remember, like sectional charts, facility directories, terminal procedures, etc., you must have current information—as in the eight-week update scheme of NOS charts—for instrument flight. Therefore, the database system used must be easily updatable. Note that updates are expensive and can run to several hundred dollars annually if every periodic update is purchased.

A very complete database, such as the *Jeppesen NavData* is available for coverage in the form of Americas or International versions and includes:

- Airports: Identifier, city/state, country, facility name, latitude/longitude, elevation, fuel service, and controlled approaches
- VORs: Identifier, city/state, country, facility name, latitude/longitude, frequencies, collocated DME (or TACAN), magnetic variation, and weather broadcast information
- NDBs: Identifier, city/state, country, facility name, latitude/longitude, frequency, and weather broadcast information
- Intersections: Identifier, country, latitude/longitude, and nearest VOR
- Communication Frequencies: Approach, arrival, control area, departure, special-use airspace (new Class B, C, and D) with altitude information; also ATIS, clearance delivery, tower, ground, unicom, and pretaxi
- Runways: Designation, length, surface, lighting, ILS/localizer frequency and ID, and pilot-controlled lighting frequency and instructions
- FSS: Identifier, reference VOR, and frequency
- MSA & MESA: Minimum safe altitude along and in proximity to a user-defined flight plan
- Map Datums: 124 predefined and 1 user-defined

TSOs and STCs

There are two TSOs that specifically involve GPS receivers at this time. These are C115a and C129.

TSO-C129 refers to the subject of integrity and RAIM, as well as databases. It was issued on December 10, 1992.

Whether all current receivers will be able to meet the stringent requirements of this TSO is unknown at this time, however, it is unlikely that lower-cost units are capable. Therefore, if you are planning to use GPS for IFR flying, you must consider whether a currently available unit might be obsolete or totally unapprovable under the TSO.

For GPS purposes, TSO-C115a applies only to equipment using multi-sensor inputs. This does include some combination LORAN-C and GPS units inside of a single housing.

Note: Approval under the older TSO-C115a does not constitute approval under TSO-C129.

STC (supplemental type certificate) covers only the physical installation of a device into a specific airplane under specified circumstances. It does not necessarily equate to the approval for use of the device.

DGPS

Currently there are problems in setting the DGPS standard for aviation differential GPS. Frequencies, locations, and even the signal content have not yet been standardized.

Beware: Some manufacturers' product literature indicate certain GPS receivers as DGPS ready. At this time the phraseology means only that they will operate with the marine RTCM differential system, that supported by the USCG, and not necessarily with any yet to be approved aviation DGPS system.

The FAA is currently testing several DGPS schemes. Only when they are finished with their research will an aviation standard be set.

Warranty and upgrade policies

Pay attention to manufacturers' stated warranty policies. You want your equipment to be covered for repairs and/or service. You also want to easily obtain the repairs and/or service, therefore determine where the repair facilities are located and whether the owner is responsible for equipment delivery under warranty repair terms.

Normally, only portable GPS receivers will require shipment to the manufacturer for service. Panel mounted equipment is usually serviceable by the selling/installing dealer.

As a further and very important point, check the manufacturer's practices and charges for updating databases.

Find out if IFR certification of equipment at a later time is possible. Be sure the unit will meet TSO-C129 if you require the GPS uses covered under it.

GPS RECEIVER FEATURES

Many choices of displays and keyboards are available. What suits you, the user, will vary from one individual to another. The single rule is that you must be satisfied with the product to use it efficiently.

The user interface

The current choices for displays on GPS receivers are LEDs (light-emitting diode), LCD (liquid crystal display), fluorescent display, and the CRT (cathode ray tube)

The LED, fluorescent, and CRT displays provide their own light source and are generally more visible than the LCD types. Manufacturers favor LCDs because of their low cost, low operating temperature, and low power drain. Some LCD displays are backlighted to aid nighttime reading. Note that almost all GPS receivers will suffer from poor readability in certain lighting conditions. Also note that panel-mounted receivers using other than LCD displays require cooling fans and use more electrical power for operation.

An additional display factor is the format and number of lines of data being shown. Displays vary from one to four lines. The dot matrix LED display provides more fully formed numbers and letters and is therefore more readable than the segmented alphanumerics usually associated with LCD displays. Further, the greater flexibility of the dot-matrix format permits an infinite variety of plain English messages using upper and lower case and special characters.

Consider your preferences regarding how data is organized and presented. Since all the data won't fit into a small display window at one time, the information is usually separated into categories, or modes, each of which may have several "pages." For example:

The nav mode will typically contain such data as your present position, range and bearing to a waypoint, desired track, cross-track error, track-angle error, ground-speed, ETE, and more.

Only two or three of these items can fit on a page at one time and still remain readable. Each manufacturer has its own ideas about which should go together and in what order. Ideally, the information you will refer to most often should be available on a single page. Subsequent information should follow on succeeding display pages in descending order of importance or use.

Keyboards generally resemble telephone touch pads for entry of alphanumeric data or a limited number of push buttons designed for picking menu selections shown on the display. Either is satisfactory, just different. In testing many GPS portable receivers I found I could get used to nearly anything with practice. With few exceptions, panel-mounted units make use of a combination of push buttons and concentric dials for menu control and display selections.

The largest caveat about all interfaces is that operation must be intuitive, not couched in obscurities. In fact, this applies to the overall receiver. It must be simple and logical to operate.

Alerts

Many of the more advanced GPS receivers can alert the user of altitude or airspace dangers based upon database information preprogrammed into the receiver. Ascertain how warning messages are handled by any receivers that interest you. Determine if they alert by audio, visual, or a combination of means. Alerts could save your life or prevent a legal problem with the FAA.

Try it first

Do not purchase any GPS receiver until you have had a chance to see it in person and actually use it. Whether you do this through a dealer or by checking with various airport buddies, make sure you have a feel for what you are spending your money on.

GPS receivers are not considered consumer items, therefore you will find very little "customer satisfaction guaranteed" phraseology in the sales literature. Be sure what you purchase will do the job you require. Beware of quickly changing technology or rules and regulations, as both are concerns regarding GPS and the FAA.

PORTABLE GPS RECEIVERS

There are several very good portable GPS receivers on the market today. Any one of them can do the job of VFR navigation for you, however, some are just plain handier than others in their physical size. If you are planning to use a portable GPS receiver mounted to the control yoke, one of the smaller and lighter units should be your choice. Watch the specifications for dimensions and weights.

WARNING: DO NOT ALLOW THE RECEIVER OR ITS WIRES TO INTERFERE WITH FLIGHT OPERATIONS!

If you are unsure how large a unit is when reading the specifications, construct a cardboard box of the same dimensions and use it as a mock up in the cockpit. You will be amazed at the handiness of some units and appalled at the sheer clumsiness of others. Of course this will depend upon your specific airplane and type of controls. After all, in a plane with a stick, there is no yoke on which to mount a portable, therefore the installation horizons are more limited.

GPS hand-helds—portables—make sense for pilots, such as renters, flying in several different airplanes. They also make sense for pilots with multiple applications such as: boating, fishing, hiking, or just plain positioning as you travel by car.

Battery life

The specified battery life shown in the various manufacturers' product literature appears to be somewhat optimistic. Experience shows battery life to be 25 to 30 percent less than that indicated.

An important factor when discussing battery life in a portable GPS receiver is the number of batteries used in the unit. The fewer batteries, the less electrical power available, and the less usable battery life. Specifically, a unit using only four batteries will have less battery life than a receiver using six batteries (providing they have the same power demands). However, the drawback of using more batteries is a larger receiver case and additional weight.

Many portable GPS receivers have a feature called *battery saver*, or similar name, that partially shuts down the receiver during intervals of non-use or for timed periods. The result is a reduction in the continuous power demand on the batteries, therefore increasing their useful life.

SPECIFICATION NOTES

The following are notes applicable to GPS receiver specifications:

GPS final accuracy is determined by the DOD selective availability (SA) degradation to 100 meters. Some accuracy specifications indicate what a unit is capable of when SA is not in use.

CDI, in the specifications, indicates a digital rendition of a graphic steering display. It is not to be confused with a real analog CDI found installed as part of a nav system.

As noted previously, databases are not all created equal. The most desirable is that which fulfills your needs, yet will not be excessively expensive to keep updated.

Interfaces of GPS with other navigational equipment and/or computers is essential for some uses. Interfacing capabilities are indicated in the specifications and must be considered if you plan on interconnecting a GPS receiver with other equipment.

The emergency-search feature found on many aviation-GPS receivers is generally a single button function displaying the bearing and distance to the nearest airport(s) or FAA facility. It is a desirable feature for emergency reasons.

The specifications that follow are based upon information supplied by the manufacturers and are subject to change without notice.

Price

Prices have not been listed with the various GPS equipment shown and described in the pages that follow, however, do appear on the comparison charts. The reason for this apparent deficiency is the very changeable aviation marketplace and the differences in profit margins between authorized dealers and nationwide discounters. Or more accurately, purchase prices will vary from full list to "street prices" generally by as much as 30 percent.

Beware, however, that many of the manufacturers have protected themselves and/or their dealers with installation requirements for panel-mounted GPS receivers. Specifically, to qualify for warranty protection you must have the installation done by a FAA-authorized entity (and be able to prove it with a Form 337) or even, in some cases, by the selling dealer. This does not apply to portable receivers.

HAND-HELD GPS RECEIVERS

Make: Collins Avionics
Model: Trooper (Fig. 10-1)
Year introduced: 1993
User interface
    LCD graphics w/CDI
    Controls: 7 buttons

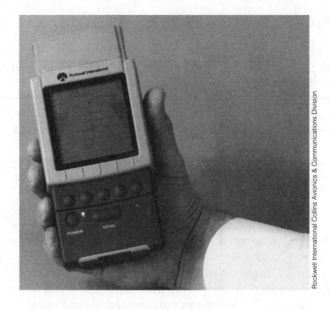

Fig. 10-1. Collins Avionics
Trooper.

Receiver
    Type: 5 channel parallel
    Satellites tracked: 5
    TTFF: 24 seconds
    Update rate: 1 second
    Accuracy (without SA)
        Position: 25 meters SEP
        Velocity: 0.1 meters/second
    Output/input: RS232
Features
    User waypoints: 100
    Emergency search: none
    Flight plans: none
    Database: none
    Map datums: 52
    Alarms: various
    Timers: event/clock
Physical attributes
    Size: 3.5 × 7.5 × 2.7 inches (whd—width, height, depth)
    Weight: 26 ounces
    Operating temperature: –10 to +60 degrees C
    Power: 6 AA/9–40 VDC
Standard components
    Manual & users guide
Optional components
    Cigarette-lighter adapter

Carrying case
NiCad battery pack/charger
External antenna
Warranty: 1 year

Make: Garmin
Model: GPS 55 AVD (Fig. 10-2)
Year introduced: 1992
User interface
    Display: LCD (3 lines) w/CDI
    Controls: backlighted alphanumeric & function
Receiver
    Type: MultiTrac
    Satellites tracked: up to 8
    TTFF: 2 minutes
    Update rate: 1 second
    Accuracy (without SA)
        Position: 15 meters
        Velocity: 0.1 knot
    Output/input: NMEA 180, 182, 183
Features
    User waypoints: 250
    Emergency search: 9 nearest facilities
    Flight plans: 9 reversible
    Database: Jeppesen (Americas or International) airports/VOR
    Map datums: 101
    Alarms: cross-track error, arrival, anchor drag

Fig 10-2. Garmin GPS 55 AVD.

Garmin International

Timers: event/clock
Physical attributes
    Size: 3.2 × 6.4 × 1.5 inches (whd) with battery pack
    Weight: 19 ounces (with battery pack)
    Operating temperature: –15 to +70 degrees C
    Case construction: Waterproof
    Power: 4 AA/5–40 VDC
    Battery life: 10 hours with battery saver mode
Standard components
    Detachable antenna w/mount
    AA battery pack
    Carrying case
    Surface mount
    Power/data cable
    Manual and user's guide
Optional components
    Cigarette-lighter adapter
Warranty: 1 year
Comments: Database updates available every 28 days.

Make: Garmin
Model: GPS 95 (Fig. 10-3)
Year introduced: 1993

Fig. 10-3. Garmin GPS 95.

User interface
    Display: dot matrix LCD w/real time moving map
    Controls: backlighted alphanumeric & function
Receiver
    Type: MultiTrac
    Satellites tracked: up to 8
    TTFF: 2 minutes
    Update rate: 1 second
    Accuracy (without SA)
        Position: 15 meters
        Velocity: 0.1 knot
        Output/input: NMEA 180, 182, 183, PC
Features
    User waypoints: 500
    Emergency search: 9 nearest facilities
    Flight plans: 20 reversible
    Database: Jeppesen (Americas or International) airports/VOR
    Map datums: 102
    Alarms: cross-track error, arrival
    Timers: event/clock
Physical attributes
    Size: 3.2 × 6.4 × 1.5 inches (whd) with battery pack
    Weight: 19 ounces (with battery pack)
    Operating temperature: –15 to +70 degrees C
    Case construction: Waterproof
    Power: 4 AA/5–40 VDC
    Battery life: 4 hours with battery saver mode
Standard components
    Detachable antenna w/mount
    Yoke mount
    AA battery pack
    Carrying case
    Surface mount
    Power/data cable
    Cigarette-lighter adapter
    Manual and user's guide
Optional components
    NiCad battery pack/charger
    PC database update kit
Warranty: 1 year
Comments: This complete unit offers real moving-map display functions, an extensive user (via PC) updatable database, and E6B functions. Manufacturer database updates are available every 28 days.

Make: ICOM
Model: GP-22 (Fig. 10-4)
Year introduced: 1993
User interface
    Display: LCD (2 lines)
    Controls: 6 buttons
Receiver
    Type: 5 channel parallel
    Satellites tracked: 5
    TTFF: 1–3 minutes
    Update rate: 1 second
    Accuracy (without SA)
        Position: 15 meters
        Velocity: 0.1 knot
    Output/input: optional
Features
    User waypoints: 99
    Emergency search: none
    Flight plans: none
    Database: none
    Alarms: not stated
    Timers: event/clock
Physical attributes
    Size: 2.6 × 5.3 × 1.5 inches (whd)
    Weight: 11.6 ounces
    Operating temperature: –10 to +50 degrees C
    Case construction: plastic
    Power: NiMH (nickel metal hydride) battery

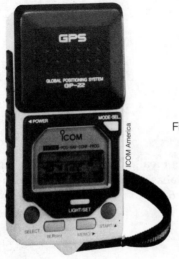

Fig. 10-4. ICOM GP-22.

Battery life: 80 minutes
Standard components
    Carrying case
    Charger
    AC adapter
    Cigarette-lighter plug/cable
    Alkaline battery pack
Optional components
    External antenna
    NMEA system support
    Mounting bracket
Warranty: 1 year

Make: Magellan
Model: NAV 5000A (Fig. 10-5)
Year introduced: 1992
User interface
    Display: LCD (4 lines) w/CDI
    Controls: alphanumeric & function
Receiver
    Type: 5 separate continuous tracking channels
    Satellites tracked: 11
    TTFF: 55 seconds
    Update rate: 1 second
    Accuracy (without SA)
        Position: 15 meters
        Velocity: 0.1 knot
    Output/input: RS232, ARGUS map compatible

Fig. 10-5. Magellan 5000A.

Features
   User waypoints: 500
   Emergency search: 5 nearest facilities
   Flight plans: 10 reversible
   Database: Jeppesen airports/VOR/NDB
   Map datums: 12
   Timers: event/clock
Physical attributes
   Size: 3.5 × 8.8 × 2.1 inches (whd)
   Weight: 30 ounces
   Operating temperature: –10 to +60 degrees C
   Case construction: splash proof
   Power: 6 AA/9–35 VDC
   Battery life: 10 hours
Standard components
   Detachable antenna w/mount
   6 foot coax antenna cable
   DC power cable
   Lanyard
   User guide
Optional components
   AC adapter
   Carrying case
   NiCad battery pack/charger
Warranty: 1 year

Make: Micrologic
Model: SuperSport GPS (Fig. 10-6)

Fig. 10-6. Micrologic SuperSport GPS.

Year introduced: 1992
User interface
    Display: LCD (4 line) w/CDI
    Controls: alphanumeric & function
Receiver
    Type: 5 channel continuous tracking
    Satellites tracked: 5
    DGPS ready: yes
    TTFF: less than 3 minutes
    Update rate: 1 second
    Accuracy (without SA)
        Position: 15 meters
        Velocity: 0.1 knot
    Output/input: NMEA 0183, 0180, RMc, PC
Features
    User waypoints: 250
    Emergency search: none
    Flight plans: 9
    Database: none
    Map datums: 134
    Alarms: arrival, cross-track error, border crossing, anchor watch, antenna cable
        short
    Timers: event/clock
Physical attributes
    Size: 3.5 × 7.6 × 2.6 inches (whd)
    Weight: 32 ounces
    Operating temperature: –20 to +70 degrees C
    Case construction: submersible waterproof
    Power: 6 AA/7–32 VDC
    Battery life: 6 hours
Standard components
    2 battery packs
    Soft carrying case
    Operator manual and guide
Optional components
    NiCad battery pack/charger
    DC power cable
    Tilt stand
    Cigarette-lighter plug
    External antenna w/25 ft cable
Warranty: 1 year
Comments: Does not provide an internal aviation database—see chapter 8 for further information. Requires an external antenna for proper performance inside an airplane.

Make: Motorola
Model: Traxar (Fig. 10-7)
Year introduced: 1992
User interface
  Display: LCD (4 line) w/CDI
  Controls: 4 fixed and 4 multi-function buttons
Receiver
  Type: 6 channel parallel tracking
  Satellites tracked: 6
  TTFF: 24 seconds
  Update rate: 1 second
  Accuracy (without SA)
    Position: 25 meters
    Velocity: 0.1 knot
  Output/input: NMEA 0183
Features
  User waypoints: 100
  Emergency search: none
  Flight plans: 10
  Database: none
  Map datums: 49
  Alarms: arrival, waypoint, cross track, low battery, position quality
  Timers: event/clock
Physical attributes
  Size: 3.6 × 7.5 × 2.0 inches (whd)
  Weight: 17 ounces
  Operating temperature: −10 to +60 degrees C
  Case construction: waterproof
  Power: 6 AA/11–34 VDC
  Battery life: 6 hours
Standard components
  Operating instructions

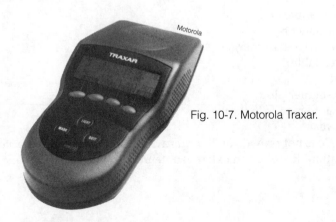

Fig. 10-7. Motorola Traxar.

Optional components
   SmartBracket (holder makes power and external antenna connections)
Warranty: 3 years
Comments: Does not provide an internal aviation database—see chapter 8 for further information. Requires an external antenna for proper performance inside an airplane.

Make: SONY
Model: IPS-360 Pyxis
Year introduced: 1991
User interface
   Display: LCD (2 line) w/modified CDI
   Controls: 14 buttons
Receiver
   Type: 4 channel parallel tracking
   Satellites tracked: 4
   TTFF: 1 to 20 minutes
   Update rate: 2 seconds
   Accuracy (without SA)
       Position: 30 meters
       Velocity: 0.3 knot
   Output/input: none
Features
   User waypoints: 100
   Emergency search: none
   Flight plans: 1
   Database: none
   Alarms: arrival
   Timers: event/clock
Physical attributes
   Size: 3.8 × 6.8 × 1.5 inches (whd)
   Weight: 21 ounces
   Operating temperature: –20 to +60 degrees C
   Power: 4 AA/12 or 24 VDC
Standard components
   Carrying case & belt
   12/24 VDC adapter
   Cigarette-lighter plug
   Mounting brackets
   23 ft antenna extension cable
Optional components
   External battery pack
Warranty: not stated
Comments: Does not provide an internal aviation database—see chapter 8 for further information. Requires an external antenna for proper performance inside an airplane.

Make: SONY
Model: IPS-760-AVD (Fig. 10-8)
Year introduced: 1993
User interface
    Display: 4.5-inch LCD (graphics) w/moving map
    Controls: 20 buttons
Receiver
    Type: 8 channel parallel
    Satellites tracked: 8
    DGPS ready: can be upgraded
    TTFF: 1 minute
    Update rate: 1 second
    Accuracy (without SA)
        Position: 9 meters
        Altitude: 25 meters
        Velocity: 0.3 knot
    Output/input: RS232, NMEA
Features
    User waypoints: 1000
    Emergency search: 10 nearest facilities
    Flight plans: 50 reversible
    Database: Jeppesen NavData
    Alarms: not stated
    Timers: event/clock
Physical attributes
    Size: 4.2 × 8.9 × 1.3 inches (whd)
    Weight: 28 ounces

Fig. 10-8. Sony PYXIS IPS-760.

Operating temperature: –20 to + 60 degrees C
Case construction: Waterproof
Power: 6 AA
Battery life: 4.5 hours
Standard components:
Carrying case
Antenna extension cable
Battery magazine
Optional components
Rechargeable battery pack
Cigarette-lighter plug
Mounting kit
Antenna extension wire
Warranty: 1 year
Comments: Database information is on user-replaceable cards. Optional, more extensive, database cards are available. Unit can hold two cards at one time.

Make: Trimble
Model: Flightmate (Fig. 10-9)
Year introduced: 1992
User interface
Display: LCD (4 lines) w/CDI
Controls: 8 multi-function buttons
Receiver
Type: 3 channel digital continuous
Satellites tracked: up to 8
TTFF: less than 2 minutes
Update rate: 1.5 seconds on external power

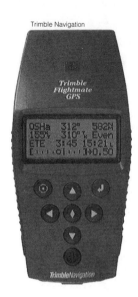

Fig. 10-9. Trimble Flightmate GPS.

Accuracy (without SA)
    Position: 15 meters
    Velocity: 0.1 knot
Output/input: none
Features
    User waypoints: 100
    Emergency search: 10 nearest facilities
    Flight plans: 2
    Database: Jeppesen (Worldwide) airports/VOR
    Map datums: 124
    Timers: event/clock
Physical attributes
    Size: 3.3 × 6.8 × 1.3 inches (whd)
    Weight: 14 ounces
    Operating temperature: 0 to + 60 degrees C
    Case construction: Plastic
    Power: 4 AA/10–32 VDC
    Battery life: 8–10 hours with battery saver mode
Standard components
    External antenna w/cable
Optional components
    Detachable antenna
    Leather carrying case
    Yoke-mounting kit
    Cigarette-lighter plug/cable
    Battery eliminator
    Manual & user's guide
Warranty: 1 year
Comments: Limited E6B functions. Built-in antenna or use external when in airplane.

Make: Trimble
Model: Flightmate PRO (Fig. 10-10)
Year introduced: 1993
User interface
    Display: LCD (4 lines) w/CDI
    Controls: 8 multi-function buttons
Receiver
    Type: 3 channel digital continuous
    Satellites tracked: up to 8
    TTFF: less than 2 minutes
    Update rate: 1.5 seconds on external power
    Accuracy (without SA)
        Position: 15 meters
        Velocity: 0.1 knot

Fig. 10-10 Trimble Flightmate GPS PRO.

Output/input: RS232
Features
    User waypoints: 100
    Emergency search: 10 nearest facilities
    Flight plans: 10
    Database: Jeppesen (Worldwide) airports/VOR/NDB
    Map datums: 124
    Timers: event/clock
Physical attributes
    Size: 3.3 × 6.8 × 1.3 inches (whd)
    Weight: 14 ounces
    Operating temperature: 0 to +60 degrees C
    Case construction: Plastic
    Power: 4 AA/10-32 VDC
    Battery life: 8-10 hours with battery-saver mode
Standard components
    External antenna w/cable
Optional components
    Detachable antenna
    Leather carrying case
    Yoke mounting kit
    Cigarette-lighter plug/cable
    Battery eliminator
    Manual & user's guide
Warranty: 1 year
Comments: Limited E6B functions. Built-in antenna or use external when in airplane. Has improved remote antenna connection. Database can be user limited to a specific region or area and queried by facility name or identifier.

PANEL-MOUNTED GPS RECEIVERS

Make: AlliedSignal (Bendix/King)
Model: KLN 90 (Fig. 10-11)
Year introduced: 1992
User interface
    Display: 3.3-inch CRT w/moving map
    Controls: 8 buttons and 2 concentric knobs
Receiver
    Type: MultiSat (single channel multi-tracking digital)
    Satellites tracked: 8
    Update rate: 1 second
    Accuracy (without SA)
        Position: 15 meters
        Velocity: 0.1 knot
    Output/input: CDI, HSI, RMI, ELT
Features
    User waypoints: 250
    Emergency search: 9 nearest airports
    Flight plans: 26 reversible
    Database: Jeppesen NavData
    Alarms: airspace & altitude
Physical attributes
    Size: 6.3 × 2 × 13.2 inches (whd)
    Weight: 6.3 pounds
    Operating temperature: –40 to +70 degrees C
    Power: 11–33 VDC
Optional components
    KCC 90 AC powered take-home case
    Upgrade to enroute and terminal capability
Warranty: 2 years

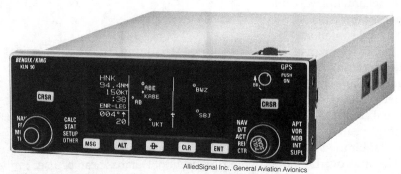

AlliedSignal Inc., General Aviation Avionics

Fig. 10-11. KLN 90.

Comments: The KLN 90 is an advanced stand-alone GPS receiver built for general aviation. Its capabilities include real moving map, complex flight calculator, a superbly complete database including airport diagrams, airport services, all facilities, all radio frequencies, altitude and airspace alerting, minimum safe altitudes, and more. The database can be updated by PC with update disks available by subscription.

Make: ARNAV
Model: Star 5000 (Fig. 10-12)
Year introduced: n/a
User interface
    Display: LED (2 line)
    Controls: 9 buttons and 1 selector knob
Receiver
    Type: 5 channel continuous parallel (Rockwell)
    Satellites tracked: 5
    TTFF: typically 1 minute
    Update rate: 1 second
    Accuracy (without SA)
        Position: 15 meters
        Altitude: 27 meters
        Velocity: 0.1 knot
    Output/input: RS232, CDI, HSI, flags, altitude, mode C XPNDR
    Antenna: patch w/preamp
Features
    User waypoints: 300
    Emergency search: 15 nearest facilities
    Flight plans: variable number
    Database: Jeppesen NavData
    Timers: event/clock
Physical attributes
    Size: 6.25 × 2 × 10.1 inches (whd)
    Weight: 3 pounds
    Operating temperature: –20 to +70 degrees C
    Power: 10–35 VDC
Warranty: 2 years
Comments: Limited E6B functions.

Fig. 10-12. ARNAV Star 5000.

Make: Ashtech
Model: AV-12 (Fig. 10-13)
Year introduced: 1993
User interface
    Display: high-resolution color CRT
    Controls: 4 buttons and 2 selector knobs
Receiver
    Type: 12 channel
    Satellites tracked: all in view
    TTFF: ‹30 seconds (with almanac)
    Update rate: 1 second
    Accuracy (without SA)
        Position: 15 meters
        Velocity: 1 cm/second
    Output/input: RS232, RS422, ARINC 429
    Antenna: external
Features
    User waypoints: 1000
    Emergency search: yes
    Flight plans: 100
    Database: Jeppesen NavData
    Timers: 2
Physical attributes
    Size: 6.25 × 2 × 13 inches (whd)
    Weight: 4.5 pounds
    Operating temperature: –20 to + 70 degrees C
    Power: 10–32 VDC
Warranty: 1 year

Make: Collins Avionics
Model: IPG-100F (Fig. 10-14)
Year introduced: 1992
User interface
    Display: LCD (4 line backlighted)
    Controls: 21 buttons
Receiver
    Type: 5 channel continuous tracking
    Satellites tracked: 5
    TTFF: ‹60 seconds
    Update rate: 1 second
    Accuracy (without SA)
        Position: 16 meters
        Velocity: 0.2 knot
    Output/input: most interfaces available

Fig. 10-13. Ashtech AV-12.

Features
　　User waypoints: 99
　　Emergency search: none
　　Flight plans: 10
　　Database: user defined
　　Map datums: 48
　　Timers: event/clock
Physical attributes
　　Size: 5.7 × 3.4 × 4.7 inches (whd)
　　Weight: 2.6 pounds
　　Operating temperature: –20 to +55 degrees C
　　Power: 28 VDC
Standard components
　　User manual and reference card
Warranty: 1 year
Options: Security module for use of full PPS and Y code functions (requires DOD user approval).

Make: Garmin
Model: GPS 100 AVD (Fig. 10-15)
Year introduced: 1991
User interface
　　Display: LCD (3 line) w/CDI
　　Controls: 21 buttons
Receiver
　　Type: MultiTrac
　　Satellites tracked: 8
　　TTFF: 2 to 2.5 minutes
　　Update rate: 1 second
　　Accuracy (without SA)
　　　　Position: 15 meters
　　　　Velocity: 0.1 knot
　　Output/input: RS232, CDI, autopilot, moving maps, RMI
Features
　　User waypoints: 100
　　Emergency search: 9 nearest airports
　　Flight plans: 10 reversible
　　Database: Jeppesen airport/VOR/NDB/OM/approach fixes

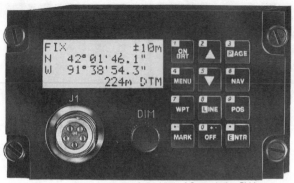

Fig. 10-14. IPG-100F GPS receiver.

Rockwell International Collins Avionics & Communications Division

Fig. 10-15. Garmin GPS 100 AVD.

Garmin International

Timers: event/clock
Physical attributes
    Size: 6.25 × 2 × 3.95 inches (whd)
    Weight: 25 oz
    Operating temperature: –15 to +70 degrees C
    Power: 6 AA/10–40 VDC
    Battery life: 8 hours
Standard components
    Blade antenna
    Aviation interface unit
    Rechargeable battery pack/charger
    User manual & reference card
Optional components
    AA battery pack
    Portable antenna
    PC software and cables
    Carrying bag
Warranty: 1 year
Comments: The database is not user updatable. However, this unit is essentially portable (battery powered) and easily transportable from one airplane to another—or anywhere for that matter.

Make: Garmin
Model: GPS 150 (Fig. 10-16)
Year introduced 1993
User interface
   Display: Fluorescent (3 line) w/CDI
   Controls: 11 buttons and 1 knob
Receiver
   Type: MultiTrac
   Satellites tracked: 8
   TTFF: 2 to 2.5 minutes
   Update rate: 1 second
   Accuracy (without SA)
      Position: 15 meters
      Velocity: 0.1 knot
   Output/input: RS232, ARINC 429, GAMA 429, NMEA 0183
Features
   User waypoints: 1000
   Emergency search: 9 nearest airports and FSS frequency
   Flight plans: 20 reversible
   Database: Jeppesen NavData
   Map datums: 124
   Alarms: arrival, proximity
   Timers: event/clock
Physical attributes
   Size: 6.25 × 2 × 5.8 inches (whd)
   Weight: 34 ounces
   Operating temperature: –15 to +70 degrees C
   Power: internal battery/10–33 VDC
   Battery life: 4 hours
Standard components
   Aviation installation kit w/antenna

Fig. 10-16. Garmin GPS 150.

Garmin International

AC charger
User manual and reference card
Optional components
PC kit
Carrying bag
Warranty: 1 year
Comments: The database is not user updatable. However, this unit is essentially portable (battery powered) and easily transportable from one airplane to another—or anywhere for that matter. Up to nine 30-item checklists can be created and kept on the 150 along with 9 programmable messages.

Make: II Morrow
Model: Flybuddy GPS 819/820 (Fig. 10-17)
Year introduced: n/a
User interface
    Display: LCD (2 line) w/CDI
    Controls: 9 buttons and 1 selector knob
Receiver
    Type: 8 channel parallel
    Satellites tracked: 8
    TTFF: 30 to 45 seconds
    Update rate: 1 second
    Accuracy (without SA)
        Position: 15 meters
        Altitude: 156 meters (with SA)
        Velocity: 0.3 knot
    Output/input: autopilot, CDI, RS232, ARGUS moving map
    Antenna: low profile w/preamp
Features
    User waypoints: 100
    Emergency search: #819—5 user waypoints
    Emergency search: #820—10 airports, 5 VORs, 5 user waypoints
    Flight plans: 10
    Database: #819—none
    Database: #820—internal worldwide airport/VOR
    Timers: event/clock
Physical attributes
    Size: 6.25 × 2 × 10.45 inches (whd)

Fig. 10-17. II Morrow Flybuddy Apollo GPS.

Weight: 3 pounds
Operating temperature: –15 to +55 degrees C
Power: 10–40 VDC
Optional components: Flybrary data card option
Warranty: 26 month
Comments: The U.S. and North American data base does not include intersections or airspace data. Regional Flybrary cards are available with concentrations of all data.

Make: II Morrow
Model: Apollo 2001 GPS (Fig. 10-18)
Year introduced: n/a
User interface
    Display: LED (3 line) w/CDI
    Controls: 8 buttons and 1 selector knob
Receiver
    Type: 8 channel parallel
    Satellites tracked: 8
    TTFF: 20 seconds w/latest ephemeris data
    Update rate: 1 second
    Accuracy (without SA)
        Position: 15 meters
        Altitude: 156 meters
        Velocity: 0.5 knot
    Output/input: RS232, serial data, fuel, moving maps, CDI, flags, autopilot
    Antenna: low profile w/preamp
Features
    User waypoints: 200
    Emergency search: nearest 20 facilities
    Flight plans: 10
    Database: Flybrary DataCard
    Alarms: airspace and altitude alert
Physical attributes
    Size: 6.25 × 2 × 10.45 inches (whd)
    Weight: 3.6 pounds
    Operating temperature: –15 to +55 degrees C
    Power: 10–40 VDC
Standard components
    power supply for use at home

Fig. 10-18. II Morrow Apollo 2001.

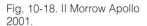
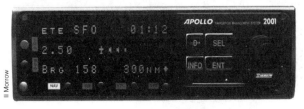

Optional components
    LORAN sensor
    fuel/air data sensor
Warranty: 26 months
Comments: The 2001 is a navigation-management system with GPS as one source
of information. Additional sources include LORAN and fuel flow/air data sensors. The database card is extensive.

Make: Magellan
Model: SkyNav 5000 (Fig. 10-19)
Year introduced: 1992
User interface
    Display: Fluorescent (2 line) w/CDI
    Controls: 10 buttons and 1 concentric knob
Receiver
    Type: 5 separate continuous tracking channels
    Satellites tracked: 11
    DGPS ready: yes
    TTFF: 55 seconds
    Update rate: 1 second
    Accuracy (without SA)
        Position: 15 meters
        Velocity: 0.1 knot
    Output/input: RS232, CDI, HSI, autopilot
Features
    User waypoints: 1000
    Emergency search: 5 nearest facilities
    Flight plans: 20
    Database: Jeppesen airports/VOR/NDB
    Map datums: 47
    Alarms: arrival
    Timers: event/clock
Physical attributes
    Size: 6.25 × 2 × 8.5 inches (whd)
    Weight: 3 pounds
    Operating temperature: –15 to +55 degrees C
    Power: 11–33 VDC

Fig. 10-19. Magellan SKYNAV
5000.

Optional components
   Jeppesen NavData card
Warranty: 1 year

Make: Narco
Model: GPS Star-Nav 900 (Fig. 10-20)
Year introduced: 1992
User interface
   Display: LED (3 lines) w/CDI
   Controls: 11 buttons and 1 selector knob
Receiver
   Type: 5 channel (independent) continuous tracking by Rockwell
   Satellites tracked: 5
   TTFF: 1 minute
   Update rate: 1 second
   Accuracy (without SA)
      Position: 15 meters
      Velocity: 0.1 knot
   Output/input: CDI, HSI, moving maps, autopilots
Features
   User waypoints: 99
   Emergency search: 10 nearest facilities
   Flight plans: 4
   Database: Jeppesen NavData
Physical attributes
   Size: 6.4 × 2.5 × 11.5 inches (whd)
   Weight: 4.5 pounds
   Power: 11–32 VDC
Optional components
   Multi-sensor DME and/or LORAN is available
Warranty: 3 years

Make: Northstar
Model: GPS-600 (Fig. 10-21)
Year introduced: n/a
User interface
   Display: LED (1 line) w/CDI

Fig. 10-20. Narco Star-Nav 900.

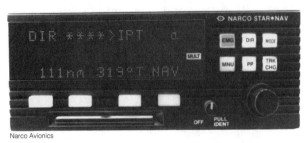

Narco Avionics

Fig. 10-21. Northstar GPS-600.

Northstar

Controls: 6 buttons and 3 concentric knobs
Receiver
    Type: 6 channel continuous tracking
    Satellites tracked: 6
    TTFF: 1 minute
    Update rate: 1 second
    Accuracy (without SA)
        Position: 15 meters
        Velocity: 0.1 knot
    Output/input: CDI, most autopilots
    Antenna: low profile w/preamp
Features
    User waypoints: 250
    Emergency search: 20 nearest facilities
    Flight plans: multiple
    Database: FliteCard (includes 2500 waypoints)
    Timers: event/clock
Physical attributes
    Size: 6.25 × 2 × 11.75 inches (whd)
    Weight: 4.5 pounds
    Operating temperature: –40 to +75 degrees C
    Power: 10–36 VDC
Warranty: 3 years
Comments: 56 day cycle database updates available.

Make: Northstar
Model: M2 Upgrade
Year introduced: n/a
User interface
    Display: LED (1 line) w/CDI
    Controls: 6 buttons and 3 concentric knobs
Receiver
    Type: 6 channel continuous tracking
    Satellites tracked: 6
    TTFF: 1 minute
    Update rate: 1 second
    Accuracy (without SA)
        Position: 15 meters

Velocity: 0.1 knot
Output/input: CDI, most autopilots
Antenna: low profile w/preamp
Features
User waypoints: 250
Emergency search: 20 nearest facilities
Flight plans: multiple
Database: FliteCard (includes 2500 waypoints)
Timers: event/clock
Warranty: 3 years
Comments: This is a dealer installed GPS package for inclusion in the model M1
LORAN-C receiver. It results in a combination GPS/LORAN unit. 56 day cycle
database updates available.

Make: Terra
Model: TGPS 400 D (Fig. 10-22)
Year introduced: 1992
User interface
Display: LCD (2 line) w/CDI
Controls: 10 buttons and 1 selector knob
Receiver
Type: 6 channel continuous parallel tracking
Satellites tracked: 8
TTFF: .5 to 3.5 minutes
Update rate: 1 second
Accuracy (without SA)
Position: 15 meters
Altitude: 35 meters
Velocity: 0.1 knot
Output/input: moving maps, fuel, CDI, EIA, RS422, autopilot
Antenna: microstrip w/preamp
Features
User waypoints: 250
Emergency search: 20 nearest facilities
Flight plans: 20
Database: Jeppesen NavData
Timers: event/clock

Fig. 10-22. Terra TGPS 400 D.

Physical attributes
    Size: 6.3 × 2 × 10.8 inches (whd)
    Weight: 2.4 pounds
    Operating temperature: −20 to +55 degrees C
    Power: 10–36 VDC
Warranty: 3 years
Comments: Limited E6B functions

Make: Trimble
Model: TNL-1000 (Fig. 10-23)
Year introduced: 1992
User interface
    Display: LCD (2 line) w/CDI
    Controls: 10 buttons and 1 selector knob
Receiver
    Type: 6 channel continuous parallel tracking
    Satellites tracked: 8
    TTFF: .5 to 3.5 minutes
    Update rate: 1 second
    Accuracy (without SA)
        Position: 15 meters
        Altitude: 35 meters
        Velocity: 0.1 knot
    Output/input: moving maps, fuel, CDI, EIA, RS422, autopilot
    Antenna: microstrip w/preamp
Features
    User waypoints: 250
    Emergency search: 20 nearest facilities
    Flight plans: 20
    Database: Jeppesen airports/VOR/NDB
    Timers: event/clock
Physical attributes
    Size: 6.3 × 2 × 10.8 inches (whd)
    Weight: 2.4 pounds
    Operating temperature: −20 to +55 degrees C
    Power: 10–36 VDC
Optional components:
    Jeppesen NavData
Warranty: 2 years

Fig. 10-23. Trimble TNL-1000.

Comments: Limited E6B functions

Make: Trimble
Model: TNL-1000DC
Year introduced: 1993
User interface
    Display: LCD (2 line) w/CDI
    Controls: 10 buttons and 1 selector knob
Receiver
    Type: 6 channel continuous parallel tracking
    Satellites tracked: 8
    TTFF: .5 to 3.5 minutes
    Update rate: 1 second
    Accuracy (without SA)
        Position: 15 meters
        Altitude: 35 meters
        Velocity: 0.1 knot
    Output/input: moving maps, fuel, CDI, EIA, RS422, autopilot
    Antenna: microstrip w/preamp
Features
    User waypoints: 250
    Emergency search: 20 nearest facilities
    Flight plans: 20
    Database: Jeppesen NavData
    Timers: event/clock
Physical attributes
    Size: 6.3 × 2 × 10.8 inches (whd)
    Weight: 2.4 pounds
    Operating temperature: –20 to +55 degrees C
    Power: 10–36 VDC
Optional components:
    FliteStar PC memory card programmer and software.
Warranty: 2 years
Comments: Limited E6B functions, dead reckoning and demo functions.

Make: Trimble
Model: TNL-2000A (Fig. 10-24)
Year introduced: 1992
User interface
    Display: LED (2 lines) w/CDI
    Controls: 9 buttons and 1 selector knob
Receiver
    Type: 6 channel continuous parallel tracking
    Satellites tracked: all-in-view
    DGPS ready: yes
    TTFF: 1.5 to 3.5 minutes

Fig. 10-24. Trimble TNL-2000A.

Trimble Navigation

Update rate: 1 second
Accuracy (without SA)
    Position: 15 meters
    Altitude: 35 meters
    Velocity: 0.1 knot
Output/input: air data computer, EIA, RS422, CDI/altitude
Antenna: microstrip w/preamp
Features
    User waypoints: 250
    Emergency search: 20 nearest airports
    Flight plans: 20
    Database: Jeppesen NavData
    Alarms: airspace
    Timers: event/clock
Physical attributes
    Size: 6.25 × 2 × 10.8 inches (whd)
    Weight: 2.75 pounds
    Operating temperature: –20 to +55 degrees C
    Power: 10–32 VDC
Warranty: 2 years

Make: Trimble
Model: TNL-2100 (Fig. 10-25)
Year introduced: 1992
User interface
    Display: LED (2 line) w/CDI
    Controls: 9 buttons and 1 selector knob
Receiver
    Type: 6 channel continuous
    Satellites tracked: all-in-view
    DGPS ready: yes
    TTFF: 1.5 to 3.5 minutes
    Update rate: 1 second
    Accuracy (without SA)
        Position: 15 meters
        Velocity: 0.1 knot
    Output/input: ADC, autopilot, CDI, RS422, altitude
    Antenna: microstrip w/preamp

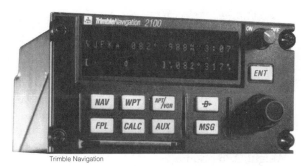

Fig. 10-25. Trimble TNL-2100.

Trimble Navigation

Features
  User waypoints: 250
  Emergency search: 20 nearest facilities
  Flight plans: 20
  Database: Jeppesen NavData
  Alarms: airspace
  Timers: event/clock
Physical attributes
  Size: 5.7 × 3 × 7.75 inches (whd)
  Weight: 2.8 pounds
  Operating temperature: –20 to +55 degrees C
  Power: 10–32 VDC
Warranty: 2 years
Comments: Dzus-type mounting used and unit is designed for use in the upper end of general-aviation aircraft.

Make: Trimble
Model: TNL-3000 GPS/LORAN (GPS specs given)
Year introduced: 1992
User interface
  Display: LED (2 lines) w/CDI
  Controls: 9 buttons and 1 selector knob
Receiver
  Type: 6 channel continuous
  Satellites tracked: 8
  DGPS ready: yes
  TTFF: 1.5 to 3.5 minutes
  Update rate: 1 second
  Accuracy (without SA)
    Position: 15 meters
    Altitude: 35 meters
    Velocity: 0.1 knot
  Output/input: EIA, RS422, CDI, altitude input, air data computer
  Antenna: microstrip w/preamp

Features
    User waypoints: 250
    Emergency search: 20 nearest airports
    Flight plans: 20
    Database: Jeppesen NavData
    Alarms: airspace
    Timers: event/clock
Physical attributes
    Size: 6.25 × 2 × 10.8 inches (whd)
    Weight: 2.75 pounds
    Operating temperature: –20 to +55 degrees C
    Power: 10–32 VDC
Warranty: 2 years
Comments: This unit is TSO (C115a multi-sensor for LORAN only) approved, others are pending.

Make: Trimble
Model: TNL-3100 GPS/LORAN (GPS specs given) (Fig. 10-26)
Year introduced: 1992
User interface
    Display: LED (2 lines) w/CDI
    Controls: 9 buttons and 2 selector knobs
Receiver
    Type: 6 channel continuous
    Satellites tracked:
    DGPS ready: yes
    TTFF: 1.5 to 3.5 minutes
    Update rate: 1 second
    Accuracy (without SA)
        Position: 15 meters
        Altitude: 35 meters
        Velocity: 0.1 knot

Fig. 10-26. Trimble TNL-3100.

Output/input: EIA, RS422, CDI, altitude input, air data computer
Antenna: microstrip w/preamp
Features
User waypoints: 250
Emergency search: 20 nearest facilities
Flight plans: 20
Database: Jeppesen NavData
Alarms: airspace
Timers: event/clock
Physical attributes
Size: 5.75 × 3 × 7.75 inches (whd)
Weight: 2.8 pounds
Operating temperature: –20 to +55 degrees C
Power: 10–32 VDC
Warranty: 2 years
Comments: This unit is TSO (C115a multi-sensor for LORAN only) approved, other TSOs are pending. Dzus-type mounting used and unit is designed for use in the upper end of general-aviation aircraft.

FAST COMPARE

To compare the features found on the various makes and models of GPS receivers you will need a side-by-side chart of similar models (Figs. 10-27 and 10-28 on pages 194 and 195).

Make	Garmin	Garmin	ICOM	Magellan	Micrologic
Model	GPS 55 AVD	GPS 95	GP-22	NAV 5000A	SuperSport
Display	LCD	LCD	LCD	LCD	LCD
Lines	3	graphic[7]	2	4	4
Receiver	Multitrac	Multitrac	5 ch	5 ch	5 ch
Satellites	8	8	5	11	5
User waypoints	250	500	99	500	250
Flight plans	9	20	none	10	9
Emergency search[1]	9 facilities	9 facilities	none	5 facilities	none
Database	apt/VOR	apt/VOR	none	apt/VOR/NDB	none
update[2]	factory	user (PC)	none	factory	none
Size (w×h×d in inches)	3.2×6.4×1.5[3]	3.2×6.4×1.5[3]	2.6×5.3×1.2	3.5×8.8×2.1	3.5×7.6×2.6
Weight	19 oz.	19 oz.	11.6 oz.	30 oz.	32 oz.
Power**	4 AA/acft	4 AA/acft	NiMH bat	6 AA/acft	6 AA/acft
Outputs	NMEA	NMEA	optional[4]	RS232	NMEA
TSO C129	no	no	no	no	no
Warranty	1 year	1 year	1 year	1 year	1 year
Price[5]	$1295	$1495	$1000	$1325	$1195
Phone	(800) 800-1020	(800) 800-1020	(206) 454-8155	(714) 394-5000	(818) 998-1216

Make	Motorola	Sony	Trimble	Trimble	Collins
Model	Traxar	IPS-760-AVD	Flightmate	Flightmate PRO	Trooper
Display	LCD	LCD	LCD	LCD	LCD
Lines	4	graphic[7]	4	4	graphic[7]
Receiver	6 ch	8 ch	3 ch	3 ch	5 ch
Satellites	6	8	8	8	5
User waypoints	100	1,000	100	100	100
Flight plans	10	50	2	10	10
Emergency search[1]	none	10 facilities	10 facilities	10 facilities	none
Database	none	Jep. NavData	apt/VOR	apt/VOR/NDB	none
update[2]	none	user (cards)	factory	factory	none
Size (w×h×d in inches)	3.6×7.5×2.0	4.2×8.9×1.3	3.3×6.8×1.3	3.3×6.8×1.3	3.5×7.5×2.7
Weight	17 oz.	28 oz.	14 oz.	14 oz.	26 oz.
Power**	6 AA/acft	6 AA	4 AA/acft	4 AA/acft	6 AA/acft
Outputs	NMEA	RS232/NMEA	none	RS232	RS232
TSO C129	no	no	no	no	no
Warranty	3 years	1 year	1 year	1 year	1 year
Price[5]	$1195	$1699	$995	$1285	$1300
Phone	(800) 272-1477	see note 6	(800) 767-8628	(800) 767-8628	(800) 321-2223

*Accuracy is not compared on this chart because it is a factor of DOD control over GPS.

**Some portable receivers can use the airplane's power system; however, they require an optional device to do so.

[1]Indicates nearest facilities (airports/VORs/NDBs) or airports only.

[2]Factory database update requires the unit be returned for service. User indicates card insert or PC update.

[3]Removable battery pack reduces size.

[4]At extra cost.

[5]Prices are factory suggested retail.

[6]Available from: Gulf Coast Avionics Corp., (813) 879-9794, or Pacific Coast Avionics Corp., (206) 931-0370.

[7]Graphic might include aviation moving map.

Fig. 10-27. Portable GPS comparison chart.

Make	AlliedSignal	ARNAV	Ashtect	Garmin	Garmin
Model	KLN 90	Star 5000	AV-12	100 AVD[3]	GPS 150[3]
Display	CRT	LED	color CRT	LCD	fluorescent
Lines	graphic	2	graphic	3	3
Receiver	MultiSat	5 ch	12 ch	MultiTrac	MultiTrac
Satellites	8	5	all-in-view	8	8
User waypoints	250	300	1,000	100	1,000
Flight plans	26	variable	100	10	20
Emergency search[1]	9 airports	15 facilities	yes	9 airports	9 airports
Database update[2]	NavData PC	NavData user (cards)	NavData user (cards)	apt/VOR/bcns factory	NavData factory
TSO C129	pending	pending	pending	no	no
Warranty	2 years	2 years	1 year	1 year	1 year
Price[4]	$6500	$4495	$8000	$3250	$3495
Phone	(913) 768-3000	(206) 847-3550	(408) 524-1400	(800) 800-1020	(800) 800-1020

Make	II Morrow	II Morrow	Magellan	Narco	Northstar
Model	820	Apollo 2001	SkyNav 5000	GPS-900	GPS-600
Display	LCD	LED	fluorescent	LED	LED
Lines	2	3	2	3	1
Receiver	8 ch	8 ch	5 ch	5 ch	6 ch
Satellites	8	8	11	5	6
User waypoints	100	200	1,000	99	250
Flight plans	10	10	20	4	multiple
Emergency search[1]	10 airports	20 facilities	5 facilities	10 facilities	20 facilities
Database update[2]	apt/VOR[6] factory/user	Flybrary Card user (cards)	apt/VOR/NDB[7] user (cards)[7]	NavData user (cards)	FliteCard user (cards)
TSO C129	no	pending	approved[8]	pending	pending
Warranty	26 months	26 months	1 year	3 years	3 years
Price[4]	$2495	$3995	$2675	$3880	$4995
Phone	(800) 525-6726	(800) 525-6726	(714) 394-5000	(800) 223-3636	(508) 897-6600

Make	Terra	Trimble	Trimble	Trimble	Trimble
Model	DGPS-400D	TNL-1000	TNL-1000DC	TNL-2000A	TNL-3000
Display	LCD	LCD	LCD	LED	LED
Lines	2	2	2	2	2
Receiver	6 ch	6 ch	6 ch	6 ch	6 ch
Satellites	8	8	8	all in view	8
User waypoints	250	250	250	250	250
Flight plans	20	20	20	20	20
Emergency search[1]	20 facilities	20 facilities	20 facilities	20 airports	20 airports
Database update[2]	NavData user (cards)	apt/VOR/NDB user (cards)	NavData user (cards)	NavData user (cards)	NavData user (cards)
TSO C129	no	no	no	pending	pending
Warranty	3 years	2 years	2 years	2 years	2 years
Price[4]	$2995	$2795	$2995	$5995	$6795
Phone	(505) 884-2321	(800) 767-8628	(800) 767-8628	(800) 767-8628	(800) 767-8628

*Accuracy is not compared on this chart because it is a factor of DOD control over GPS.

[1]Indicates nearest facilities (airports/VORs/NDBs) or airports only.

[2]Factory database update requires the unit be returned for service. User indicates card insert or PC update.

[3]Can be operated on batteries and/or not physically mounted to the panel.

[4]Prices are factory suggested retail.

[5]Search of user entered waypoints only.

[6]Optional Flybrary card available.

[7]NavData optional.

[8]TSO approval might not include all requirements.

Fig. 10-28. Panel-mounted GPS comparison chart.

A

Abbreviations

THE FOLLOWING LIST contains the abbreviations most common to GPS and aviation navigation:

1 PPS—1 pulse per second
1 PPM—1 pulse per minute
AC—advisory circular
A/D—analog to digital
ADAM—Airport Datum Monument Program
ADF—automatic direction finder
ADS—automatic dependent surveillance
AE—antenna electronics
AFB—Air Force Base
AFS—Air Force Station
AGL—above ground level
AHRS—Attitude and Heading Reference System
AIMS—Airspace Traffic Control Radar Beacon System IFF System
ARTCC—Air Route Traffic Control Center
AS—antispoofing
ATC—Air Traffic Control
ATIS—Automated Terminal Information System
ATMS—Advanced Traffic Management System
BCD—binary code decimal
BPSK—bi-phase shift keying
C/A—coarse/acquisition
C/A-code—coarse/acquisition code
CCW—coded continuous wave
CDI—course deviation indicator
CEP—circular error probable
CGS—Civil GPS Service
CMOS—complementary metal oxide semiconductor
CNI/NAV—communications, navigation, and identification/navigation

C/No—carrier-to-noise ratio
CNS—communication, navigation, and surveillance
CONUS—continental United States
CRPA—controlled radiation pattern antenna
CS—control segment
CSE—course selection error
CW—continuous wave
DAC—digital-to-analog converter
dB—decibel ($X = 10 \log \times dB$)
DGPS—Differential Global Positioning System
DH—decision height
DMA—Defense Mapping Agency
DME—distance measuring equipment
DME/P—precision distance measuring equipment
DOD—Department of Defense
DOP—dilution of precision
DOT—Department of Transportation
DR—dead reckoning
drms—distance root mean squared
DRS—Dead Reckoning System
ECBF—earth-centered body fixed
ECCM—electronic counter-countermeasures
ECEF—earth centered earth fixed
EDM—electronic distance measurement
EFIS—Electronic Flight Instrument System
EHF—extremely high frequency
EM—electromagnetic
EMI—electromagnetic interference
EMP—electromagnetic pulse
ESGN—electrically suspended gyro navigator
FAA—Federal Aviation Administration
FAATC—Federal Aviation Administration Technical Center
FAF—final approach fix
FAR—Federal Aviation Regulation
FCC—Federal Communications Commission
FL—flight level
FM—frequency modulation
FMS—Flight Management System
FOC—full operational capability
FOM—figure of merit
FRP—Federal Radionavigation Plan
FRPA—fixed radiation pattern antenna
FRPA-GP—FRPA ground plane
FSD—full-scale development
FTE—flight technical error
FTMI—Flight Operations and Air Traffic Management Integration

GA—general aviation
GaAs—gallium arsenide
GCA—ground control approach
GDOP—geometric dilution of precision
GHz—gigahertz
GIB—GPS integrity broadcast
GLONASS—Global Navigation Satellite System (CIS system)
GMT—Greenwich Mean Time
GPS—Global Positioning System
GPSIC—GPS Information Center
HDOP—horizontal dilution of precision
HF—high frequency
HOW—hand over word
HQ USAF—Headquarters U.S. Air Force
HSI—horizontal situation indicator
HV—host vehicle
Hz—hertz (cycles per second)
ICAO—International Civil Aviation Organization
ICNS—Integrated Communication, Navigation, and Surveillance
IF—intermediate frequency
IFF—Identification Friend or Foe
IFR—instrument flight rules
ILS—instrument landing system
INS—inertial navigation system
IOC—initial operational capability
IOT&E—initial operational test and evaluation
J/S—jamming-to-signal ratio
kHz—kilohertz
L1—GPS primary frequency, 1575.42 MHz
L2—GPS secondary frequency, 1227.6 MHz
LADGPS—Local Area Differential GPS
LEP—linear error probable
LF—low frequency
LOFF—LORAN flight following
LOP—line of position
LORAN—long-range navigation
MAP—missed approach point
MCS—Master Control Station
MCT—mean corrective maintenance time
MCW—modulated carrier wave
MDA—minimum descent altitude
MF—medium frequency
MHz—megahertz
MLS—Microwave Landing System
mm—millimeters
MNP—Master Navigation Plan

MNPS—Minimum Navigational Performance Specifications
MOPS—Minimum Operational Performance Standard
M/S—meters per second
MSL—mean sea level
MTBF—mean time between failures
MTBM—mean time between maintenance
MTTR—mean time to repair
N/A—not applicable
NAS—National Airspace System
NASA—National Aeronautics and Space Administration
NAV-msg—navigation message
NDB—nondirectional beacon
nm—nautical mile
NNSS—Navy Navigation Satellite System (Transit)
NOAA—National Oceanic and Atmospheric Administration
NOS—National Ocean Service
NOTAM—notice to airmen
NPN—National Plan for Navigation
NRL—Naval Research Laboratory
ns—nanosecond
NSA—National Security Agency
OBS—omni bearing select
OC—obstruction clearance
Omega—ground-based VLF navigation system (not an acronym)
PAR—precision approach radar
PC—personal computer
P-Code—precise code
PDOP—position dilution of precision
POS/NAV—positioning and navigation
PPM—parts per million (10)
PPS—Precise Positioning Service
PPS-SM—PPS security module
PRC—pseudorange corrections
PRN—pseudorandom noise
PTTI—precise time and time interval
PVT—position, velocity, and time
R&D—research and development
R&E—research and engineering
RACON—radar transponder beacon
RAIM—receiver autonomous integrity monitoring
RAM—reliability and maintainability
RBN—radiobeacon
RCVR—receiver
RDF—radio direction finder
R,E&D—research, engineering and development
RF—radio frequency

RFI—radio frequency interference

RMS—root mean square

RNAV—area navigation (radio)

RNP—required navigation performance

RRC—range-rate corrections

RSS—Root Sum Square

RT—remote terminal

RTCA—Radio Technical Commission for Aeronautics

RVR—runway visual range

S/A or **SA**—selective availability

SAMSO—Space and Missile Systems Organization

SAR—search and rescue

SEP—spherical error probable

SHF—super high frequency

SPS—Standard Positioning Service

STC—supplemental type certification

STOL—short takeoff and landing

SV—space vehicle

TACAN—tactical air navigation

TAI—International Atomic Time

TC—type certification

TD—time difference

TDOP—time dilution of precision

TERPS—terminal instrument procedures

TFOM—time figure of merit

TOA—time of arrival

Transit—satellite-based navigation system (not an acronym)

TSO—Technical Standard Order

TTFF—time to first fix

TVOR—terminal VOR

UE—user equipment

UERE—user equivalent range error

UHF—ultra high frequency

URE—user range error

USAF—United States Air Force

USCG—United States Coast Guard

USGS—United States Geological Survey

USNO—United States Naval Observatory

UTC—Universal Time Coordinated

VDOP—vertical dilution of precision

VFR—visual flight rules

VHF—very high frequency

VLBI—very long baseline interferometry

VLF—very low frequency

VNAV—vertical navigation

VOR—very high frequency omnidirectional range

VORTAC—collocated VOR and TACAN
VTOL—vertical takeoff and landing
WADGPS—Wide Area Differential GPS
WARC—World Administrative Radio Conference
WGS—World Geodetic System
WGS-84—World Geodetic System—1984

B

Chart reference systems

THE FOLLOWING INFORMATION about map datums and their relation to maps and charts is quoted from the Federal Radionavigation Plan of 1992. Of particular importance is the section on Aeronautical Charts:

Geodetic datums are basic control networks used to establish the precise geographic position and elevation of features on the surface of the earth. They are established at all levels of government—international, national, and local—and form the legal basis for all positioning and navigation. Within the last 20 years, there have been great advances in our knowledge of the shape and size of the earth (i.e., our geodetic knowledge). The old datums are no longer scientifically relevant (although otherwise still relevant). In recent years, geodesy and navigation tended toward Earth Centered Body Fixed (ECBF) coordinate systems. These are Cartesian coordinate systems with origins at the center of mass of the earth and whose axes rotate with the earth. The old datums have generally been based on localized surface monumentations (and associated agreements) and defined by a reference ellipsoid that was not earth-centered.

The Department of Defense (DOD) Global Positioning System (GPS) is based on the World Geodetic System of 1984 (WGS 84). WGS 84 is an ECBF coordinate system upon which all U.S. military and much civilian navigation, geodesy, and survey will be based. Within the U.S., the National Geodetic Survey (NGS) is the primary civilian legal authority for the establishment of U.S. datums. Until recently, the datum used throughout most of the U.S. and Canada was the North American Datum of 1927 (NAD 27). NAD 27 is a surface or horizontal datum. Until recently, nearly all nautical charts, aeronautical charts, Federal surveys, and associated data provided by the National Ocean Service (NOS) of the National Oceanic and Atmospheric Administration (NOAA) were legally established with respect to NAD 27. In 1986, NGS completed a new horizontal datum known as the North American Datum of 1983 (NAD 83) which, for purposes of navigation and relative survey, is effectively equivalent to WGS 84. Although NAD 27 is still heavily used, increasingly datum products and activities are being converted to NAD 83.

There is also a vertical (i.e., height) datum. Until recently, there has been the National Geodetic Vertical Datum of 1929 (NGVD 29). In 1991, the NGS completed the North American Vertical Datum of 1988 (NAVD 88). Vertical datum products and activities have begun the conversion from NGVD 29 to NAVD 88. The conversion between GPS determined heights—i.e., ellipsoid heights—and vertical datum—i.e., orthometric—heights is made by using a geoid model associated with the respective vertical datum. NGS has developed a geoid model, GEOID 90, to support such conversions.

Aeronautical charts

The ultimate responsibility for the accuracy of air cartographic positional data rests with NOS. Section 307(b)(3) of the Federal Aviation Act authorizes the FAA, subject to available appropriations, to arrange for the publication of aeronautical maps and charts necessary for the safe and efficient movement of aircraft in air navigation utilizing the facilities and assistance of other Federal agencies. NOS, in turn, performs many of these services. Within the National Airspace System (NAS), the NGS establishes the basic U.S. datum that legally controls all positioning with the U.S. The Nautical Charting Division (NCD) of NOS conducts the Airport Obstruction Clearance Surveys (OC Surveys) which establish the positioning for 750 U.S. major civil airports and all navigational aids to existing U.S. datums. The NGS has completed the Airport Datum Monument Program (ADAM) which established datum monuments on 1400 non-OC surveyed airports. The ADAM data, which include end-of-runway coordinates, were determined using GPS and are available in NAD 27 and NAD 83 datums. The FAA began the conversion from NAD 27 to NAD 83 on October 15, 1992. The Aeronautical Charting Division verifies all other positions before they are charted.

The FAA conversion from NAD 27 to NAD 83 has a major impact on FAA. All positional data currently used within the NAS will require conversion. The NGS has determined that the horizontal differences between NAD 27 and NAD 83 are as large as 450 meters in Hawaii, 160 meters in Alaska, and 100 meters in the central U.S. The horizontal differences are not uniformly distributed. Vertical datum differences are relatively minor and transformation will be performed after horizontal datum conversion. The new NAD 83 coordinate system will be, for all practical purposes, identical to the WGS 84 employed by DOD for GPS and inertial navigation systems.

C

Datums used
around the world

THE POSITIONS OF land masses, navigation points, etc., are determined by coordinates from an established mapping system called a *datum*. Around the world there are many datums in use.

The GPS is designed around the World Geodetic Survey 1984 (WGS-84). It is convertible to and from nearly all other datums, however, to apply the fix coordinates upon one datum-base map to another datum-based map can cause errors of thousands of feet. Note that some U.S. maps still reference the North American 1927 Datum.

Most GPS receivers are capable of converting from WGS-84 to other datums. This allows immediate application of the PVT information displayed. The various datums and coordinate systems found around the world include:

ADINDAN—Ethiopia, Mali, Senegal, Sudan
AFGOOYE—Somalia
AIN EL ABD 1970—Bahrain Island, Saudi Arabia
ANNA 1 ASTRO 1965—Cocos Islands
ARC 1950—Botswana, Lesotho, Malawi, Swaziland, Zaire, Zambia, Zimbabwe
ARC 1960—Kenya, Tanzania
ASCENSION ISLAND 1958—Ascension Island
ASTRO B4 SOROL ATOLL—Tern Island
ASTRO BEACON "E"—Iwo Jima Island
ASTRO DOS 71/4—St. Helena Island
ASTRONOMIC STATION 1952—Marcus Island
AUSTRALIAN GEODETIC 1966—Australia, Tasmania Island
AUSTRALIAN GEODETIC 1984—Australia, Tasmania Island
BELLEVUE (IGN)—Efate and Erromango Islands
BERMUDA 1957—Bermuda Islands
BOGOTA OBSERVATORY—Colombia

CAMPO INCHAUSPE—Argentina
CANTON ASTRO 1966—Phoenix Islands
CAPE—South Africa
CAPE CANAVERAL—Florida, Bahama Islands
CARTHAGE—Tunisia
CHATHAM 1971—Chatham Island (New Zealand)
CHUA ASTRO—Paraguay
CORREGO ALEGRE—Brazil
DJAKARTA (BATAVIA)—Sumatra Island (Indonesia)
DOS 1968—Gizo Island (New Georgia Islands)
EASTER ISLAND 1967—Easter Island
EUROPEAN 1950—Austria, Belgium, Denmark, Finland, France, Germany, Gibraltar, Greece, Italy, Luxembourg, Netherlands, Norway, Portugal, Spain, Sweden, Switzerland
EUROPEAN 1950—Egypt
EUROPEAN 1950—Iran
EUROPEAN 1950—Sicily
EUROPEAN 1979—Austria, Finland, Netherlands, Norway, Spain, Sweden, Switzerland
FINLAND HAYFORD 1910—Finland
GANDAJIKA BASE—Republic of Maldives
GEODETIC DATUM 1949—New Zealand
GUAM 1963—Guam Island
GUX 1 ASTRO—Guadalcanal Island
HJORSEY 1955—Iceland
HONG KONG 1963—Hong Kong
INDIAN (BLDS)—Bangladesh, India, Nepal
INDIAN (TLND)—Thailand, Vietnam
IRELAND 1965—Ireland
ISTS 073 ASTRO 1969—Diego Garcia
JOHNSTON ISLAND 1961—Johnston Island
KAN DAWALA—Sri Lanka
KERGUELEN ISLAND—Kerguelen Island
KERTAU 1948—West Malaysia, Singapore
L.C. 5 ASTRO—Cayman Brac Island
LIBERIA 1964—Liberia
MAHE 1971—Mahe Island
MARCO ASTRO—Salvage Islands
MASSAWA—Eritrea (Ethiopia)
MERCHICH—Morocco
MIDWAY ASTRO 1961—Midway Island
MILITARY GRID REFERENCE SYSTEM (MGRS)—U.S.A.
MINNA—Nigeria
MASIRAH NAHRWAN—Masirah Island (Oman)
NAHRWAN—United Arab Emirates
NAHRWAN—Saudi Arabia

NAPARIMA, BWI—Trinidad and Tobago
NORTH AMERICAN 1927 (NAD-27)—Mean Value (CONUS)
NORTH AMERICAN 1927—Alaska
NORTH AMERICAN 1927—Bahamas (excluding; Salvador Island)
NORTH AMERICAN 1927—San Salvador Island
NORTH AMERICAN 1927—Canada (including Newfoundland Islands)
NORTH AMERICAN 1927—Canal Zone
NORTH AMERICAN 1927—Caribbean (Barbados, Caicos Islands, Cuba, Dominican Republic, Grand Cayman, Jamaica, Leeward Islands, Turks Island)
NORTH AMERICAN 1927—Central America (Belize, Costa Rica, El Salvadore, Guatemala, Honduras, Nicaragua)
NORTH AMERICAN 1927—Cuba
NORTH AMERICAN 1927—Greenland (Hayes Peninsula)
NORTH AMERICAN 1927—Mexico
NORTH AMERICAN 1983—Alaska, Canada, Central America, CONUS, Mexico
OBSERVATORIO 1966—Corvo and Flores Islands (Azores)
OLD EGYPTIAN—Egypt
OLD HAWAIIAN—Mean Value
OMAN—Oman
ORDNANCE SURVEY OF GREAT BRITAIN 1936—England, Isle of Man, Scotland, Shetland Islands, Wales
PHILIPPINES LUZON—Philippines (excluding Mindanao Island)
PICO DE LAS NIEVES—Canary Islands
PITCAIRN ASTRO 1967—Pitcairn Island
PROVISIONAL SOUTH AMERICAN 1956—Bolivia, Chile, Colombia, Ecuador, Guyana, Peru, Venezuela
PROVISIONAL SOUTH CHILEAN 1963—South Chile
PUERTO RICO—Puerto Rico, Virgin Islands
QATAR NATIONAL—Qatar
QORNOQ—South Greenland
REUNION—Mascarene Island
ROME 1940—Sardinia Island
SANTO (DOS)—Espirito Santo Island
SAO BRAZ—Sao Miguel, Santa Maria Islands (Azores)
SAPPER HILL 1943—East Falkland Island
SCHWARZECK—Namibia
SOUTH AMERICAN 1969—Argentina, Bolivia, Brazil, Chile, Colombia, Ecuador, Guyana, Paraguay, Peru, Venezuela, Trinidad and Tobago
SOUTH ASIA—Singapore
SOUTHEAST BASE—Porto Santo and Madeira Island
SOUTHWEST BASE—Faial, Graciosa, Pico, Sao Jorc and Terceira Islands (Azores)
SWEDEN—RT90
TIMBALA 1948—Brunei and East Malaysia (Sarawak and Sabah)
TOKYO—Japan, Korea, Okinawa
TRISTAN ASTRO 1968—Tristan da Cunha

UNIVERSAL TRANSVERSE MERCATOR (UTM) VITI LEVU 1916—Viti Levu
Island (Fiji Islands)
WAKE-ENIWETOK 1960—Marshall Islands
WORLD GEODETIC SYSTEM 1972—WGS-72
WORLD GEODETIC SYSTEM 1984—WGS-84
ZANDERIJ—Surinam

D

Federal policy and plans

Excerpts from the *Federal Policy and Plans For the Future Radionavigation System Mix 1992* showing the planned end-of-use dates for many navaid systems currently in use by general aviation:

The Federal Government operates radionavigation systems as one of the necessary elements to enable safe transportation and encourage commerce within the United States. It is a goal of the Government to provide this service in a cost-effective manner. In order to meet both civil and military radionavigation needs, the Government has established a series of radionavigation systems over a period of years. Each system utilizes the latest technology available at the time it was introduced to meet existing or unfulfilled needs. This statement addresses how and for what period each system should be part of the Federal radionavigation systems mix.

The Department of Defense is deploying a new high-technology radionavigation system, the Global Positioning System (GPS), which will have wide civil application on a global basis. This system has the potential to meet or better the accuracy and coverage capabilities of most other radionavigation systems. Consequently, if the full civil potential of GPS is realized, the Department of Transportation will consider phasing out some of the existing radionavigation systems.

Any decision to discontinue Federal operation of existing systems will depend upon many factors including: (a) resolution of GPS accuracy, coverage, integrity, and financial issues; (b) determination that the systems mix meets civil and military needs currently met by existing systems; (c) availability of civil user equipment at prices that would be economically acceptable to the civil community; (d) establishment of a transition period of 1,015 years; and (e) resolution of international commitments.

Radionavigation systems operated by the U.S. Government will be available subject to direction by the National Command Authority (NCA) because of a real or potential threat of war or impairment to national security. Radionavigation systems will be operated as long as the U.S. and its allies accrue greater military benefit than do adversaries. Operating agencies may cease operations or change characteristics and signal formats of radionavigation systems during a dire national emergency.

Individual system plans

LORAN-C. LORAN-C is the Federally provided radionavigation system for the U.S. Coastal Confluence Zone (CCZ). It provides navigation, location, and timing services for both civil and military air, land and marine users. LORAN-C is approved as a supplemental air navigation system. It is also approved for nonprecision approaches at certain airports. The LORAN-C system now serves the 48 conterminous states, their coastal areas, and certain parts of Alaska. It is expected to remain part of the radionavigation mix through the year 2015.

The DOD requirement for the LORAN-C system will end December 31, 1994. Operations conducted by the United States Coast Guard at overseas stations will be phased out by the end of 1994. In the case of stations located outside the U.S., discussions continue between the U.S. and the respective foreign governments concerning the continuation of service after the DOD requirement terminates.

VOR/DME. VOR/DME provides users with the primary means of air navigation in the National Airspace System (NAS). VOR/DME, as the international standard for civil air navigation in controlled airspace, will remain a short-range aviation navigation system through the year 2010.

The DOD requirement for and use of VOR/DME will terminate when aircraft are properly integrated with GPS and when GPS is certified to meet RNP for national and international controlled airspace. The target date is the year 2000.

TACAN. TACAN is a short-range navigation system used primarily by military aircraft.

The DOD requirement for and use of land-based TACAN will terminate when aircraft are properly integrated with GPS and when GPS is certified to meet RNP in national and international controlled airspace. The target date is the year 2000. The requirement for shipboard TACAN will continue until a suitable replacement is operational.

ILS, MLS. ILS is the standard civil landing system in the U.S. and abroad, and is protected by ICAO (International Civil Aviation Organization) agreement to January 1, 1998. ICAO has selected the MLS as the international standard precision approach system, with implementation targeted for 1998. MLS is expected to gradually replace ILS in national and international civil aviation. The FAA and DOD plan to have MLS collocated with ILS to minimize the transition impact.

Radiobeacons. Maritime and aeronautical radiobeacons serve the civilian user community with low-cost navigation. Some maritime radiobeacons will be modified to carry differential GPS correction signals. This may cause errors when used by certain aeronautical receivers; therefore, these maritime radiobeacons should not be used for aviation. Aeronautical radiobeacons and maritime radiobeacons, which will carry differential GPS correction signals, will remain part of the radionavigation systems mix into the next century. Many of the remaining maritime radiobeacons may be phased out after the year 2000.

GPS. GPS is a DOD-developed, worldwide, satellite-based radionavigation system that will be the DOD's primary radionavigation system well into the next century. The operational capability of GPS is of significant interest to both civil and military users. The term Full Operational Capability (FOC) is of particular significance to the Department of Defense as it defines the condition when full and supportable military capability is provided by a system. GPS FOC will be declared by the Secretary of Defense when 24 operational (Block II/IIA) satellites are operating in their assigned orbits and when the constellation has successfully completed testing for operational military functionality. An Initial Operational Capability (IOC) will be attained when 24 GPS satellites (Block I/II/IIA) are operating in their assigned orbits, are available for navigation use and can provide levels of service as specified below. Notification of IOC by the Secretary of Defense to the Secretary of Transportation will follow an assessment by the Air Force, as the system operator, that the constellation can sustain the stated levels of accuracy and availability throughout the IOC period. IOC is planned to occur in mid-1993 and military FOC is planned in 1995.

Prior to IOC, GPS is considered a developmental system. System operation, including signal availability and accuracy, is subject to change at the discretion of DOD. Operations conducted using the developmental system are, therefore, subject to disruption if it is necessary to adjust system operating parameters in support of system testing. At IOC, the GPS will have achieved its earliest operational configuration and the Standard Positioning Service (SPS) will be available as specified below.

Subsequent to IOC, any planned disruption of the SPS in peacetime will be subject to a minimum of 48-hour advance notice provided by the DOD to the Coast Guard GPS Information Center (GPSIC) and the FAA Notice to Airmen (NOTAM) system. A disruption is defined as periods in which the GPS is not capable of providing SPS as specified below. Unplanned system outages resulting from system malfunctions or unscheduled maintenance will be announced by the GPSIC and NOTAM systems as they become known. The Coast Guard and the FAA will notify civil users when the GPS is approved for navigation.

Differential GPS. Differential GPS (DGPS) is a system in which differences between observed and predicted GPS signals at a particular location are transmitted to users as a differential correction to upgrade the precision and performance of the user's receiver processor. Several DOT agencies are planning to provide DGPS services.

Aeronautical DGPS. The FAA, in cooperation with DOD, is planning to use differential corrections to GPS/SPS in the provision of RNP in the National Airspace System, including approaches to landing in all weather conditions.

All licensed communication links, including those used to transmit differential GPS corrections, are subject to the direction of the NCA. DOD/DOT will not constrain the use of SPS-based differential GPS service as long as applicable U.S. statutes and international agreements are adhered to.

E

Phase I IFR approval

ON JUNE 9, 1993, the FAA announced implementation of Phase I of GPS IFR flight. Due to its historical and technical importance it is quoted:

INFORMATION: IFR APPROVAL OF GLOBAL POSITIONING SYSTEM (GPS) OPERATIONS

From: Director, Flight Standards Service, AFS-1
          Director, Aircraft Certification Service, AIR-1
To: All Regional Flight Standards Division Managers
          All Certification Directorates
          Director of Aviation System Standards, AVN-1
          All FSDO, AEG, and ACO Managers

This memorandum provides approval for U.S. civil operators to use GPS equipment to conduct oceanic, domestic enroute, and terminal IFR operations.

This memorandum also provides approval for U.S. civil operators to use GPS to fly certain FAR Part 97 VOR, VOR/DME, NDB, NDB/DME, TACAN, and RNAV instrument approach procedures (IAP), and special IAP's, under the conditions specified herein. All nonprecision IAP's established by the U.S. military are also included except those noted: "NOT FOR CIVIL USE." All requests to ATC and all clearances from ATC related to these nonprecision approach procedures must specify the procedure by its published title.

All GPS IFR approvals are restricted to the flight phases, provisions, and conditions described in this memorandum and its attachments and the applicable AFM or AFM supplement. GPS systems used for the approach overlay program must meet the requirements of TSO C-129. The equipment must be installed in accordance with appropriate airworthiness installation requirements. Acceptable methods of installation are provided in the attached Notices N8110.47 and N8110.48 dated April 23.

Certain limited IFR operations can also be conducted with GPS equipment which is installed and operated in accordance with the AFS/AIR joint interim guidance memorandum, dated July 20, 1992, Interim Guidance for Installation and Approval of Global Positioning System (GPS) Equipment in Aircraft.

Note: It should be noted that equipment approved to TSO C-115a do not meet the requirements of TSO C-129.

GPS IFR operations in oceanic areas can be conducted as soon as the proper avionics is installed provided the provisions and limitations of this memorandum and the FAA-approved AFM or AFM supplement, as appropriate, are met. Prior to initial operational capability (IOC) of GPS by the Department of Defense (DOD), aircraft using GPS for oceanic IFR operations must also be equipped with other approved means, such as dual INS or dual Omega, that are approved and appropriate for the intended route of flight. These systems must be installed and operating and be actively monitored by the flight crew. Upon IOC, a GPS installation with TSO C-129 authorization in class A1, A2, B1, B2, C1, or C2 may be used to replace one of the other approved means of long-range navigation. A single GPS installation with these classes of equipment which provides RAIM for integrity monitoring may also be used on short oceanic routes which have only required one means of long-range navigation. All GPS oceanic IFR operations must be conducted in accordance with Attachment 1.

For GPS domestic enroute and terminal IFR operations, the VOR, DME, TACAN, and/or NDB avionics necessary to receive all of the ground-based facilities appropriate for the route to the destination airport and any required alternate airport must be installed and operational. The ground-based facilities necessary for these routes must also be operational. All GPS enroute and terminal IFR operations must be conducted in accordance with Attachment 1.

For GPS nonprecision approach operations, the VOR, DME, TACAN, and/or NDB avionics necessary to receive all of the ground based facilities which define the instrument approaches to be used at all required alternate airports must be installed and operational. The ground-based facilities which support these approaches must also be operational. The ground-based and airborne equipment requirements for the destination airport must be in accordance with Attachment 2.

Air carrier and commercial operators conducting GPS IFR operations shall meet the appropriate provisions of their approved operations specifications. Additional guidance for approval of operations specifications will be forthcoming in Order 8400.10 and an interim handbook bulletin. Advisory circulars also will be published to provide more detailed information regarding the various implementation phases, procedures for approval, and use of GPS navigation systems.

Thomas C. Accardi   Thomas E. McSweeny

ATTACHMENT 1

Use of GPS for IFR Oceanic, Domestic Enroute, and Terminal Area Operations

1. Prior to DOD declaration of initial operational capability, (IOC), GPS equipment can be used to conduct IFR operations in oceanic, domestic enroute, and terminal areas if the provisions and limitations of this memorandum and this attachment are met. This approval permits the use of GPS in a manner that is con-

sistent with current navigation requirements, provided the following restrictions are complied with.

a. The GPS navigation equipment used must be approved in accordance with the requirements specified in TSO C-129 and the installation must be made in accordance with Notice 8110.47 or 8110.48 or the AFS/AIR joint guidance memorandum dated July 20, 1992.

b. The basic integrity for these operations must be provided by receiver autonomous integrity monitoring (RAIM) or an equivalent method.

c. The GPS operation must be conducted in accordance with the FAA-approved flight manual or flight manual supplement.

d. Aircraft using GPS equipment under IFR must be equipped with an approved and operational alternate means of navigation appropriate to the route to be flown. This traditional navigation equipment must be actively used by the flight crew to monitor the performance of the GPS system.

e. Procedures must be established for use in the event that significant GPS navigation outages are predicted to occur. In situations where this is encountered, the flight must rely on other approved equipment, delay departure, or cancel the flight.

f. Aircraft navigating by GPS are considered to be RNAV aircraft. Therefore, the appropriate equipment suffix must be included in the ATC flight plan.

2. After DOD declaration of IOC, GPS equipment can continue to be used as described in paragraph 1 above. However, active monitoring of the traditional navigation equipment, as required by paragraph 1d above, is not required unless the installation uses RAIM for integrity monitoring. For these systems, active monitoring by the flight crew is required when the RAIM capability of the GPS equipment is lost.

ATTACHMENT 2

GPS Nonprecision Approach Implementation Provisions

1. General. Civil aircraft are not authorized to use GPS to fly any segment of any instrument approach under IFR weather conditions unless the provisions and limitations in this memorandum and its attachments are met. The following general provisions apply to all IFR GPS instrument approach operations.

a. The GPS avionics used to fly any nonprecision instrument approach must meet TSO C-129 requirements for instrument approach operations or equivalent criteria. The GPS installation must be made in accordance with Notice 8110.47 or 8110.48. The avionics database must contain the nonprecision approaches to be flown. All associated databases and charted GPS instrument flight procedures used must contain coordinates relative to the North American Datum 1983 (NAD 83) or World Geodetic System of 1984 (WGS 84).

Note: GPS equipment installed and operated in accordance with the AFS/AIR guidance memorandum dated July 20, 1992, may continue to be used for RNAV approaches. This equipment can also be used to conduct nonprecision approaches under the provisions specified for Phase I of the "overlay" program. These systems are not approved for "overlay" program Phase II or Phase III.

b. An approach cannot be flown unless that instrument approach is retrieved from the avionics database. The GPS avionics must store the location of all waypoints, intersections, and/or navigation aids required to define the approach and present them in the order as depicted on the published nonprecision instrument approach procedure chart. Approaches must be flown in accordance with the FAA-approved flight manual or flight manual supplement.

c. Any required alternate airport must have an approved instrument approach procedure, other than GPS or LORAN-C, which is anticipated to be operational at the estimated arrival time.

d. The general approval to use GPS to fly instrument approaches is initially limited to U.S. airspace. The use of GPS in any other airspace must be expressly authorized by the Administrator. GPS instrument approach operations outside the United States must also be authorized by the appropriate sovereign authority.

e. Procedures must be established for use in the event that GPS outages are predicted to occur. In situations where this occurs, the flight must rely on other approved equipment, delay departure, or cancel the flight.

f. The general approval to use GPS to fly instrument approaches is initially limited to VOR, VOR/DME, NDB, NDB/DME, TACAN, and RNAV procedures. The use of GPS for any other instrument approach must be expressly authorized by the Administrator.

2. Nonprecision Approach "Overlay" Program Phases. The nonprecision approach "overlay" program consists of three phases. The basic integrity for these operations must be provided by receiver autonomous integrity monitoring (RAIM) or an equivalent method.

a. Phase I (prior to IOC). GPS can be used as the primary IFR flight guidance during an instrument approach, prior to the DOD declaring initial operational capability (IOC) for the GPS constellation, if the following additional Phase I provisions and limitations are met.

(1) Additional integrity must be provided by active monitoring of the approach using the traditional ground-based navaid(s) required by that published approach (e.g., VOR, NDB, RNAV). An appropriate non-GPS RNAV system must be used to actively monitor approach progress if GPS is used to execute an RNAV instrument approach.

(2) The appropriate ground navaid(s) which define the published approach being flown and the traditional user avionics required to fly that approach must be operating.

(3) The approach must be requested and approved by its current published

name. Modification of the published instrument approach name is not required for Phase I of the nonprecision approach "overlay" program.

b. Phase II (after IOC). After the DOD declares IOC for the GPS constellation, GPS can be used as the primary IFR flight guidance during an instrument approach without actively monitoring the underlying navaid(s) which define the approach being used. The following provisions and limitations must be met.

(1) The ground-based navaid(s) required for the published approach must be operating and the user avionics for the approach must be installed and operational but need not be operating during the approach if RAIM provides integrity for the approach navigation data. For systems that do not use RAIM for integrity, the ground-based navaid(s) and the airborne avionics needed to provide the equivalent integrity must be installed and operating during the approach. GPS equipment installed and operated in accordance with the AFS/AIR guidance memorandum dated July 20, 1992, is not approved for "overlay" program Phase II.

(2) The approach must be requested and approved by its published name, i.e., NDB RWY 24, VOR 24, etc. Modification of the published instrument approach name is not required for Phase II of the "overlay" program.

c. Phase III (after name modification). Phase III of the nonprecision approach "overlay" program shall not start until the DOD has declared IOC for the GPS constellation and the instrument approach procedure has been modified to include "GPS" in the title of the approach. Systems that do not use RAIM to provide integrity for the approach navigation data must continue to be operated as described in paragraph 2b above. Systems that use RAIM to provide integrity for the navigation data used during the approach may be operated as follows:

(1) During Phase III, the ground-based navaid(s) that traditionally defined the published approach at the destination airport may be inoperative. The traditional airborne equipment required for the published approach(s), other than the GPS approach, at the destination airport need not be installed.

(2) The ground-based navaid(s) required for the published approach at any required alternate airport must be operational. The airborne equipment required for the published approach, other than GPS, at any required alternate airport must also be operational. The avionics required to receive the standard ICAO navaid(s) that define the route to be flown from the departure to the destination and the route to any required alternate airport must be installed and operational. GPS equipment installed and operated in accordance with the AFS/AIR guidance memorandum dated July 20, 1992, is not approved for "overlay" program Phase III.

(3) The published approach must be identified as a GPS approach (e.g., "VOR/DME or GPS RWY 24").

(4) The GPS approach to be flown must be requested by its published name (e.g., "GPS RWY 24").

F

GLONASS

GLONASS (Global Navigation Satellite System) is the future Russian satellite worldwide navigation system, expected to be operational in the mid 1990s. As with GPS, it is to provide PVT (position, velocity, time) information. Under development since the 1970s, it was first officially announced in February 1982 when the then Soviet Union informed the International Telecommunication Union of its intention to establish a global satellite navigation system.

The first GLONASS satellite was launched in late October 1982, with twice-yearly subsequent launches since. GLONASS satellites are launched three at a time from the Soviet Space Center in central Asia. Unfortunately, most of the satellites have been plagued with very short life spans, typically functioning for less than two years each.

Initially, official information about launches and the GLONASS system development were scarce, generally, only a few lines in Pravda mentioning an obscure space launch. In the late 1980s the Soviets began to promote the system and officially release information. Today the United States and former Soviet Union are parties to an agreement concerning the common use of GPS and GLONASS for civil aircraft navigation.

It is interesting to note that the GLONASS was designed and implemented along similar lines as GPS, however, public proclamation states that the design is for civilian marine and aviation interests. Claims have been made about not having anything similar to GPS's Selective Availability feature, however, this is not confirmed.

At least one American company has designed a positioning receiver/processor to utilize both GPS and GLONASS satellites.

Technical

The GLONASS satellites transmit spread-spectrum signals similar to GPS signals. Both systems transmit L-band coarse/acquisition code (C/A-code) signals:

GLONASS—a single pseudorandom number (PRN) code is used by all satellites and each GLONASS satellite transmits on its own individual frequency.

GPS—a single L1 transmit frequency is used by all satellites and each satellite transmits a different PRN code.

GLONASS specifications

The operational GLONASS constellation will have 24 satellites in 3 planes inclined 64.3 degrees to the equator. Each of the three planes will contain 8 satellites evenly separated from adjacent satellites by 45 degrees. The orbit of GLONASS satellites is nominally circular with an altitude of 19,150 km and a period of 11 hours 15 minutes and 44 seconds.

Constellation:
24 satellites (3 planes of 8 each)

Satellite orbits:
Altitude: 19,100 kilometers
Orbit time: 11 hours 15 minutes 44 seconds
Inclination: 64.3 degrees
Separation: 45 degrees

Frequency:
1602 MHz + 0.5625 k (k indicates the satellite number)

Accuracy:
Position: 100 meters 95%
Time: 1 microsecond

G

GPS aviation equipment suppliers

The following businesses supply GPS equipment and/or supplies for GPS users in general aviation:

AeroNautical Products
P.O. Box 90349
San Diego, CA 92169
(800) 801-6187
(619) 270-4111
FAX (619) 270-4044

AlliedSignal, Inc.
General Aviation Avionics
400 North Rogers Rd.
Olathe, KS 66062
(913) 768-3000

Air Chart Systems
13368 Beach Ave.
Venice, CA 90292
(310) 822-1996
FAX (310) 822-3268

ARNAV Systems, Inc.
P.O. Box 73730
Puyallup, WA 98373
(206) 847-3550
FAX (206) 847-3966

Ashtech, Inc.
1170 Kifer Rd.
Sunnyvale, CA 94086
(408) 524-1400
FAX (408) 524-1500

Bendix/King
see AlliedSignal, Inc., General Aviation Avionics

Celestaire, Inc.
416 South Pershing
Wichita, KS 67218

Collins Avionics & Communications Division
350 Collins Rd. NE
Cedar Rapids, IA 52498
(319) 395-2208

Eventide Avionics
Division of Eventide, Inc.
One Alsan Way
Little Ferry, NJ 07643
(800) 446-7878
(201) 641-1200
FAX (201) 641-1640

Garmin International
9875 Widmer Rd.
Lenexa, KS 66215
(913) 599-1515
(800) 800-1020

ICOM America, Inc.
2380 116th Ave., NE
Belleview, WA 98004
(206) 454-8155
FAX (206) 454-1509

II Morrow, Inc.
2345 Turner Rd. SE
Salem, OR 97302
(800) 525-6726

J&W Marketing Associates, Inc. (TELDIX)
P.O. Box 2131
Westerly, RI 02891
(800) 833-2131
(401) 322-8020
FAX (401) 322-8021

Jeppesen Sanderson
55 Inverness Dr. East
Englewood, CO 80112
(303) 799-9090

Magellan Systems Corp.
960 Overland Ct.
San Dimas, CA 91773
(714) 394-5000
FAX (909) 394-7050

Magnavox Electronic Systems
2829 Maricopa St.
Torrance, CA 90503
(310) 618-7026

Memtec Corp.
19B Keewaydin Dr.
Salem, NH 03079
(603) 893-8080
FAX (603) 893-8699

MentorPlus Software, Inc.
P.O. Box 356
22775 Airport Rd.
NE Aurora, OR 97002-0356
(800) 628-4640
(503) 678-1431
FAX (503) 678-1480

Micrologic
9610 De Soto Ave.
Chatsworth, CA 91311
(818) 998-1216
FAX (818) 998-3658

Mid-Continent Instruments
7706 E. Osie
Wichita, KS 67207
(800) 821-1212
(316) 683-5619
FAX (316) 683-1861

Motorola, Inc.
4000 Commercial Ave.
Northbrook, IL 60062
(800) 272-1477

Narco Avionics, Inc.
270 Commerce Dr.
Fort Washington, PA 19034
(800) 223-3636
(215) 643-2905
FAX (215) 643-0197

Northstar Technologies, Inc.
30 Sudbury Rd.
Acton, MA 01720
(508) 897-6600

OACCQPOINT
2925 California St.
Torrance, CA 90503
(310) 618-7076

Peacock Systems, Inc.
Suite 325, Civil Air Terminal
Hanscom Field
Bedford, MA 01730
(800) 533-1012
(617) 274-8218
FAX (617) 274-8130

Rockwell International
see Collins Avionics & Communications Division

S-Tec Corp.
946 Pegram
Mineral Wells, TX 76067
(800) USA-STEC
(817) 325-9406
FAX (817) 325-3904

SONY Corp. of America
Sony Dr.
Park Ridge, NJ 07656
(201) 930-7416
FAX (201) 930-7179

Terra Corp.
3520 Pan Am Freeway NE
Albuquerque, NM 87107
(505) 884-2321
FAX (505) 884-2384

Trimble Navigation Limited
2105 Donley Dr.
Austin, TX 78758
(800) 767-8628
FAX (512) 836-9413

ZYCOM Corp.
18 Loblolly Ln.
Wayland, MA 01778
(800) 955-6466
(508) 358-5052

Glossary

These definitions are quoted from the Federal Radionavigation Plan of 1992.

accuracy The degree of conformance between the estimated or measured position and/or velocity of a platform at a given time and its true position or velocity. Radionavigation-system accuracy is usually presented as a statistical measure of system error and is specified as:

> *predictable* The accuracy of a radionavigation system's position solution with respect to the charted solution. Both the position solution and the chart must be based upon the same geodetic datum. (Note: appendix C discusses chart-reference systems and the risks inherent in using charts in conjunction with radionavigation systems.)

> *repeatable* The accuracy with which a user can return to a position whose coordinates have been measured at a previous time with the same navigation system.

> *relative* The accuracy with which a user can measure position relative to that of another user of the same navigation system at the same time.

air traffic control (ATC) A service operated by appropriate authority to promote the safe, orderly, and expeditious flow of air traffic.

approach reference datum A point at a specified height above the runway centerline and the threshold. The height of the MLS approach reference datum is 15 meters (50 ft). A tolerance of plus 3 meters (10 feet) is permitted.

area navigation (RNAV) A method of navigation that permits aircraft operations on any desired course within the coverage of station-referenced navigation signals or within the limits of self-contained system capability.

automatic dependent surveillance A function in which aircraft automatically transmit navigation data derived from onboard navigation systems via a data link for use by air traffic control.

availability The availability of a navigation system is the percentage of time that the services of the system are usable. Availability is an indication of the ability of the system to provide usable service within the specified coverage area. Signal availability is the percentage of time that navigational signals transmitted from external sources are available for use. Availability is a function of both the physical characteristics of the environment and the technical capabilities of the transmitter facilities.

block II/IIA The satellites that will form the initial GPS constellation at FOC.

circular error probable (CEP) In a circular normal distribution (the magnitudes of the two one-dimensional input errors are equal and the angle of cut is 90°), circular error probable is the radius of the circle containing 50% of the individual measurements being made, or the radius of the circle inside of which there is a 50% probability of being located.

common-use systems Systems used by both civil and military sectors.

conterminous U.S Forty-eight adjoining states and the District of Columbia.

coordinate conversion The act of changing the coordinate values from one system to another; e.g., from geodetic coordinates (latitude and longitude) to Universal Transverse Mercator grid coordinates.

coordinated universal time (UTC) UTC, an atomic time scale, is the basis for civil time. It is occasionally adjusted by one-second increments to ensure that the difference between the uniform time scale, defined by atomic clocks, does not differ from the earth's rotation by more than 0.9 seconds.

coverage The coverage provided by a radionavigation system is that surface area or space volume in which the signals are adequate to permit the user to determine position to a specified level of accuracy. Coverage is influenced by system geometry, signal power levels, receiver sensitivity, atmospheric noise conditions, and other factors which affect signal availability.

differential A technique used to improve radionavigation system accuracy by determining positioning error at a known location and subsequently transmitting the determined error, or corrective factors, to users of the same radionavigation system, operating in the same area.

distance root mean square (drms) The root-mean-square value of the distances from the true location point of the position fixes in a collection of measurements. As used in this document, 2 drms is the radius of a circle that contains at least 95% of all possible fixes that can be obtained with a system at any one place. Actually, the percentage of fixes contained within 2 drms varies between approximately 95.5% and 98.2%, depending on the degree of ellipticity of the error distribution.

enroute A phase of navigation covering operations between a point of departure and termination of a mission. For airborne missions the enroute phase of navigation has two subcategories, enroute domestic and enroute oceanic.

enroute domestic The phase of flight between departure and arrival terminal phases, with departure and arrival points within the conterminous United States.

enroute oceanic The phase of flight between the departure and arrival terminal phases, with an extended flight path over an ocean.

flight technical error (FTE) The contribution of the pilot in using the presented information to control aircraft position.

full operational capability (FOC) For GPS, this is defined as the capability that will occur when 24 operational (Block II/IIA) satellites are operating in their assigned orbits and have been tested for military functionality and meet military requirements.

geocentric   Relative to the earth as a center, measured from the center of mass of the earth.

geodesy   The science related to the determination of the size and shape of the earth by such direct measurements as triangulation, leveling, and gravimetric observations; which determines the external gravitational field of the earth and, to a limited degree, the internal structure.

geometric dilution of precision (GDOP)   GDOP is all geometric factors that degrade the accuracy of position fixes derived from externally referenced navigation systems.

inclination   One of the orbital elements (parameters) that specifies the orientation of an orbit. Inclination is the angle between the orbital plane and a reference plane, the plane of the celestial equator for geocentric orbits and the ecliptic for heliocentric orbits.

initial operational capability (IOC)   For GPS, this is defined as the capability that will occur when 24 GPS satellites (Block I/II/IIA) are operating in their assigned orbits and are available for navigation use.

integrity   Integrity is the ability of a system to provide timely warnings to users when the system should not be used for navigation.

meaconing   A technique of manipulating radio frequency signals to provide false navigation information.

nanosecond (ns)   One billionth of a second.

national airspace system (NAS)   The NAS includes U.S. airspace; air navigation facilities, equipment and services; airports or landing areas; aeronautical charts, information and service; rules, regulations and procedures; technical information; and labor and material used to control and/or manage flight activities in airspace under the jurisdiction of the United States. System components shared jointly with the military are included.

national command authority (NCA)   The NCA is the President or the Secretary of Defense, with the approval of the President. The term NCA is used to signify constitutional authority to direct the Armed Forces in their execution of military action. Both movement of troops and execution of military action must be directed by the NCA; by law, no one else in the chain of command has the authority to take such action.

nautical mile (nm)   A unit of distance used principally in navigation. The International Nautical Mile is 1852 meters long.

navigation   The process of planning, recording, and controlling the movement of a craft or vehicle from one place to another.

nonprecision approach   A standard instrument approach procedure in which no electronic glide slope is provided (e.g., VOR, TACAN, LORAN-C, or NDB).

precise time   A time requirement accurate to within 10 milliseconds.

precision approach   A standard instrument approach procedure using a ground-based system in which an electronic glide slope is provided (e.g., ILS).

radiodetermination The determination of position, or the obtaining of information relating to positions, by means of the propagation properties of radio waves.

radiolocation Radiodetermination used for purposes other than those of radionavigation.

radionavigation The determination of position, or the obtaining of information relating to position, for the purposes of navigation by means of the propagation properties of radio waves.

reliability The probability of performing a specified function without failure under given conditions for a specified period of time.

required navigation performance A statement of the navigation performance accuracy necessary for operation within a defined airspace, including the operating parameters of the navigation systems used within that airspace.

rho (ranging mode) A mode of operation of a radionavigation system in which the times for the radio signals to travel from each transmitting station to the receiver are measured rather than their differences (as in the hyperbolic mode).

sigma *See* Standard Deviation.

spherical error probable (SEP) The radius of a sphere within which there is a 50% probability of locating a point or being located. SEP is the three-dimensional analogue of CEP.

standard deviation (sigma) A measure of the dispersion of random errors about the mean value. If a large number of measurements or observations of the same quantity are made, the standard deviation is the square root of the sum of the squares of deviations from the mean value divided by the number of observations less one.

supplemental air navigation system An approved navigation system that can be used in controlled airspace of the National Airspace System in conjunction with a primary means of navigation.

surveillance The observation of an area or space for the purpose of determining the position and movements of craft or vehicles in that area or space.

terminal A phase of navigation covering operations required to initiate or terminate a planned mission or function at appropriate facilities. For airborne missions, the terminal phase is used to describe airspace in which approach control service or airport traffic control service is provided.

terminal area A general term used to describe airspace in which approach control service or airport traffic control service is provided.

theta Bearing or direction to a fixed point to define a line of position.

time interval The duration of a segment of time without reference to where the time interval begins or ends.

world geodetic system (WGS) A consistent set of parameters describing the size and shape of the earth, the positions of a network of points with respect to the center of mass of the earth, transformations from major geodetic datums, and the potential of the earth (usually in terms of harmonic coefficients).

Postscript

While writing this book, I have often paused to muse about some of the dilemmas that surround GPS.

Hand-held calculators were the craze of the early 1970s. They went from several hundred dollars for very capable devices to the hundred dollar area for a simple unit. By the 1980s, hand-held calculators could be found in blister packs at department store check-outs and a few were even given away as solar powered calculator/business cards. I predict that GPS receivers will be in the same check-outs by the year 2005 and be loaded with specialized databases including one for highway travel/vacation information.

Of interest, the smallest GPS receiver built to date is the Rockwell *NavCor MicroTracker,* measuring a diminutive $2 \times 2.8 \times .53$ inches. No doubt small sizes will aid in the proliferation of GPS.

With the great miracle of GPS and its ease of use, I really wonder what will happen in the event of total and permanent failure a number of years down the road. Will the users still possess (if they ever did) the skills to manually position themselves? Will they position themselves without GPS any more than today's high school students can do math without a calculator?

Will GPS become everything it is supposed to? You bet—and more than you can imagine. In this vein, Patrick Navin, a columnist for *Pipers Magazine,* made a very appropriate remark regarding the naysayers of GPS. He said, "Pay them no mind—their ancestors thought the earth was flat."

Index